Answer Key

Algebra I
Common Core Regents
Course Workbook

Donny Brusca

Answer Key

Algebra I
Common Core Regents
Course Workbook

Donny Brusca

ISBN 978-1500722302

www.CourseWorkbooks.com

Table of Contents

PRACTICE PROBLEMS 4
REGENTS QUESTIONS 59
SAMPLE REGENTS EXAM 83

Practice Problems

II. LINEAR EQUATIONS AND INEQUALITIES

Properties ~ Page 31

1. $-\frac{2}{3}$	2. $\frac{3}{2}$
3. $(2\div 1)\neq(1\div 2)$ or any similar counterexample.	4. No. For example, when we subtract the whole number 5 from the whole number 2, the result is –3, which is *not* a whole number.
5. commutative property of multiplication	6. associative property of addition
7. ab	8. $-ab$
9. No. For example, when we divide the integer 1 by the integer 2, the result is $\frac{1}{2}$, which is *not* an integer.	10. If $\frac{a}{b}$ and $\frac{c}{d}$ are rational numbers and a, b, c, and d are *non-zero* integers, then $\frac{a}{b}\div\frac{c}{d}=\frac{a}{b}\times\frac{d}{c}=\frac{ad}{bc}$. Since the set of integers is closed under multiplication, ad and bc are integers, so $\frac{ad}{bc}$ is rational.
11. $5x+25$	12. $4b-16$
13. $-2x+2$	14. $-3a+3b$
15. $-1-y$	16. $a+1$
17. $rs+rt=r(s+t)$	18. $2x+10=2(x+5)$

Solving Linear Equations in One Variable ~ Page 37

1. $3(m-2)=18$ $3m-6=18$ $3m=24$ $m=8$	2. $4n-n=-12$ $3n=-12$ $n=-4$
3. $-5=-(y+1)-y$ $-5=-y-1-y$ $-5=-2y-1$ $-4=-2y$ $2=y$	4. $15x-3(3x+4)=6$ $15x-9x-12=6$ $6x-12=6$ $6x=18$ $x=3$
5. $3x+8=5x$ $8=2x$ $4=x$	6. $8p+2=4p-10$ $4p+2=-10$ $4p=-12$ $p=-3$
7. $5(2x-7)=15x-10$ $10x-35=15x-10$ $-35=5x-10$ $-25=5x$ $-5=x$	8. $5(x-2)=2(10+x)$ $5x-10=20+2x$ $3x-10=20$ $3x=30$ $x=10$
9. $-4(y-3)=5(2y-6)$ $-4y+12=10y-30$ $12=14y-30$ $42=14y$ $3=y$	10. $3(x-2)-2(x+1)=5(x-4)$ $3x-6-2x-2=5x-20$ $x-8=5x-20$ $-8=4x-20$ $12=4x$ $3=x$

Solving Linear Inequalities in One Variable ~ Page 42

1. $x \le 4$

2. $x > -1$

3.
$$2x - 5 \le 11$$
$$2x \le 16$$
$$x \le 8$$

4.
$$-6y + 1 > 25$$
$$-6y > 24$$
$$y < -4$$

5.
$$-4 > 2(r - 3)$$
$$-4 > 2r - 6$$
$$2 > 2r$$
$$1 > r$$
$$r < 1$$

6.
$$-\tfrac{4}{3}(x - 3) \ge 12$$
$$-\tfrac{4}{3}x + 4 \ge 12$$
$$-\tfrac{4}{3}x \ge 8$$
$$-4x \ge 24$$
$$x \le -6$$

7.
$$-4(2m - 6) + m > 3m + 4$$
$$-8m + 24 + m > 3m + 4$$
$$-7m + 24 > 3m + 4$$
$$24 > 10m + 4$$
$$20 > 10m$$
$$2 > m$$
$$m < 2$$

8.
$$-5(p + 1) \ge -p + 11$$
$$-5p - 5 \ge -p + 11$$
$$-5 \ge 4p + 11$$
$$-16 \ge 4p$$
$$-4 \ge p$$
$$p \le -4$$

9. $-4 \le x < 2$

10. $-2 < x \le 3$

11. $x < -1$ or $x > 4$

12.

13.
$$3 \le 2x + 1 < 9$$
$$2 \le 2x < 8$$
$$1 \le x < 4$$

14.
$$-2 < 3x + 4 \le 10$$
$$-6 < 3x \le 6$$
$$-2 < x \le 2$$

There are 2 positive integer solutions, $\{1, 2\}$

Solving Proportions by Linear Equations ~ Page 48

1.	2.
$\frac{2}{x+1}=\frac{5}{15}$ $5(x+1)=30$ $5x+5=30$ $5x=25$ $x=5$	$\frac{p-5}{4}=\frac{p+6}{5}$ $4(p+6)=5(p-5)$ $4p+24=5p-25$ $24=p-25$ $49=p$
3.	4.
$\frac{7y-5}{3}=\frac{9y}{4}$ $27y=4(7y-5)$ $27y=28y-20$ $-y=-20$ $y=20$	$\frac{1-h}{5}=\frac{h-4}{-2}$ $5(h-4)=-2(1-h)$ $5h-20=-2+2h$ $3h-20=-2$ $3h=18$ $h=6$

Solving Equations with Fractions ~ Page 50

1. $\frac{x}{16}+\frac{1}{4}=\frac{1}{2}$ $16\left(\frac{x}{16}\right)+16\left(\frac{1}{4}\right)=16\left(\frac{1}{2}\right)$ $x+4=8$ $x=4$	2. $\frac{x}{2}+\frac{x}{6}=2$ $6\left(\frac{x}{2}\right)+6\left(\frac{x}{6}\right)=6(2)$ $3x+x=12$ $4x=12$ $x=3$
3. $\frac{3}{5}x+\frac{2}{5}=4$ $5\left(\frac{3}{5}x\right)+5\left(\frac{2}{5}\right)=5(4)$ $3x+2=20$ $3x=18$ $x=6$	4. $\frac{3}{4}x+2=\frac{5}{4}x-6$ $4\left(\frac{3}{4}x\right)+4(2)=4\left(\frac{5}{4}x\right)+4(-6)$ $3x+8=5x-24$ $8=2x-24$ $32=2x$ $16=x$
5. $\frac{3}{4}x=\frac{1}{3}x+5$ $12\left(\frac{3}{4}x\right)=12\left(\frac{1}{3}x\right)+12(5)$ $9x=4x+60$ $5x=60$ $x=12$	6. $\frac{3}{4}(x+3)=9$ $\frac{3}{4}x+\frac{9}{4}=9$ $4\left(\frac{3}{4}x\right)+4\left(\frac{9}{4}\right)=4(9)$ $3x+9=36$ $3x=27$ $x=9$
7. $\frac{1}{2}(18-5x)=\frac{1}{3}(6-4x)$ $9-\frac{5}{2}x=2-\frac{4}{3}x$ $6(9)-6\left(\frac{5}{2}x\right)=6(2)-6\left(\frac{4}{3}x\right)$ $54-15x=12-8x$ $54=12+7x$ $42=7x$ $6=x$	8. $\frac{2}{3}\left(2x-\frac{1}{2}\right)=13$ $\frac{4}{3}x-\frac{2}{6}=13$ $6\left(\frac{4}{3}x\right)-6\left(\frac{2}{6}\right)=6(13)$ $8x-2=78$ $8x=80$ $x=10$

Literal Equations ~ Page 54

1.	2.
$2m+2p=16$ $2p=-2m+16$ $p=-m+8$	$bx-2=K$ $bx=K+2$ $x=\frac{K+2}{b}$
3.	4.
$c=2m+d$ $c-d=2m$ $\frac{c-d}{2}=m$	$bx-3a=c$ $bx=3a+c$ $x=\frac{3a+c}{b}$
5.	6.
$V=lwh$ $\frac{V}{lh}=w$	$A=\frac{bh}{2}$ $2A=bh$ $\frac{2A}{b}=h$
7.	8.
$3x-ax=b$ $x(3-a)=b$ $x=\frac{b}{3-a}$	$2ax=-bx+1$ $2ax+bx=1$ $x(2a+b)=1$ $x=\frac{1}{2a+b}$

III. VERBAL PROBLEMS

Translating Expressions ~ Page 59

1. $7x-5$	2. $2(x-8)$
3. $33-g$	4. $20-2d$
5. $4x+10$	6. $\frac{n}{12}$
7. $y+y+1+y+2+y+3=4y+6$	8. $x+3+x+5+x+7=3x+15$
9. $t=$ Tommy's age $t-4=$ Donny's age $t-4+7=t+3=$ Camille's age Sum is $3t-1$	10. $h=$ horse's lifespan $h+70=$ stork's lifespan $4(h+70)=4h+280=$ whale's lifespan Sum is $6h+350$

Translating "Each" ~ Page 67

1. $80x+75$	2. $5y+100$
3. $20-0.50g$	4. $30,000-2000m$

Translating Equations ~ Page 69

1. $2(3x+2)=22$	2. $0.30(n+4)+0.50n=3.60$
3. $7x$ deer, $3x$ foxes $3x=210$ $x=70$ So, $7x=490$ deer	4. $7x$ boys, $10x$ girls $7x+10x=357$ $17x=357$ $x=21$ So, $7x=147$ boys

Translating Inequalities ~ Page 72

1. $b+b+9<144$ $2b+9<144$	2. $0.75a+b\leq 100$

Word Problems – Linear Equations ~ Page 75

1. $a+a+5=19$ $2a+5=19$ $2a=14$ $a=7 \quad a+5=12$ Jamie is 12 years old.	2. $c+2c=561$ $3c=561$ $c=187 \quad 2c=374$ There are 187 crickets and 374 grasshoppers.
3. $c+3c=20$ $4c=20$ $c=5 \quad 3c=15$ There were 15 robins.	4. $f+2f+4=16$ $3f+4=16$ $3f=12$ $f=4 \quad 2f+4=12$ There are 4 freshmen and 12 sophomores.
5. $4m-8=28$ $4m=36$ $m=9$ Minnie owns 9 video discs.	6. $4(m+100)+12m=3056$ $4m+400+12m=3056$ $16m+400=3056$ $16m=2656 \quad 4(m+100)$ $m=166 \quad =1064$ There were 1064 balcony tickets sold.
7. $0.10(3n)+0.25(n+4)+0.05n=4.60$ $0.3n+0.25n+1+0.05n=4.60$ $0.6n+1=4.60$ $0.6n=3.60$ $n=6$ 6 nickels, 18 dimes, 10 quarters	8. $6.50s+9.00(150-s)=1180.00$ $6.5s+1350-9s=1180$ $-2.5s+1350=1180$ $-2.5s=-170$ $s=68$ 68 small and 82 large

Word Problems – Inequalities ~ Page 79

1. $2n-5>23$ $2n>28$ $n>14$ Smallest integer is 15.	2. $n+7n\leq 60$ $8n\leq 60$ $n\leq 7.5$ Largest two integers are 7 and 49.
3. $5.95h\geq 215$ $h\geq 36.1344...$ He needs to work 37 hours.	4. $6n>3n+30$ $3n>30$ $n>10$ They need to make 11 toys.
5. Convert \$1.50 per 30 mins. to \$3/hr. $3(h-1)+5\leq 12.50$ $3h-3+5\leq 12.50$ $3h+2\leq 12.50$ $3h\leq 10.50$ $h\leq 3.5$ She can park 3.5 hours.	6. $2n-(150+0.50n)\geq 500$ $2n-150-0.50n\geq 500$ $1.5n-150\geq 500$ $1.5n\geq 650$ $n\geq 433\frac{1}{3}$ They must sell 434 programs.

IV. LINEAR GRAPHS

Determining Whether a Point is on a Line ~ Page 82

1. Yes. $7 = 3(3) - 2$? $7 = 9 - 2$? $7 = 7$ ✓	2. No. $9 = \frac{1}{2}(4) + 5$? $9 = 2 + 5$? $9 \neq 7$
3. Yes. $0 = 4(0)$? $0 = 0$ ✓	4. Yes. $2(-4) - 3(-2) = -2$? $-8 + 6 = -2$? $-2 = -2$ ✓
5. No. $4(-4) - (3) = -13$? $-16 - 3 = -13$? $-19 \neq -13$	6. Yes. $5(-2) - 2(-4) = -2$? $-10 + 8 = -2$? $-2 = -2$ ✓
7. No. $2(-5) - (-1) = -11$? $-10 + 1 = -11$? $-9 \neq -11$	8. Yes. $4(3) = 3(-2) + 18$? $12 = -6 + 18$? $12 = 12$ ✓
9. $2x + 6(-2) = 4$ $2x - 12 = 4$ $2x = 16$ $x = 8$	10. $4k + (3) = -9$ $4k = -12$ $k = -3$

Lines Parallel to Axes ~ Page 85

1. $x=9$	2. $y=1$
3. $x=0$	4. $y=0$
5. (5,0)	6. 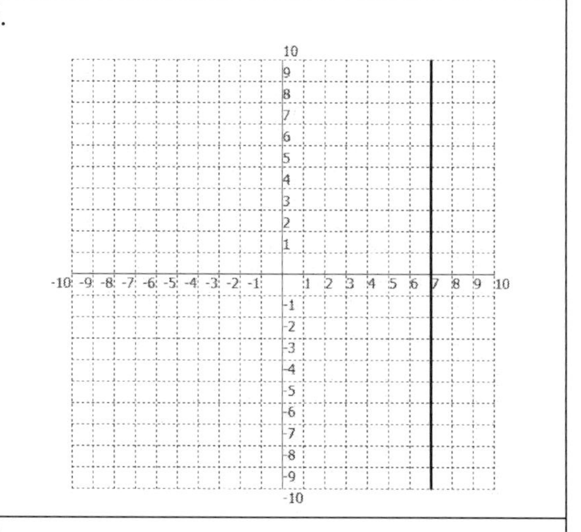
7. 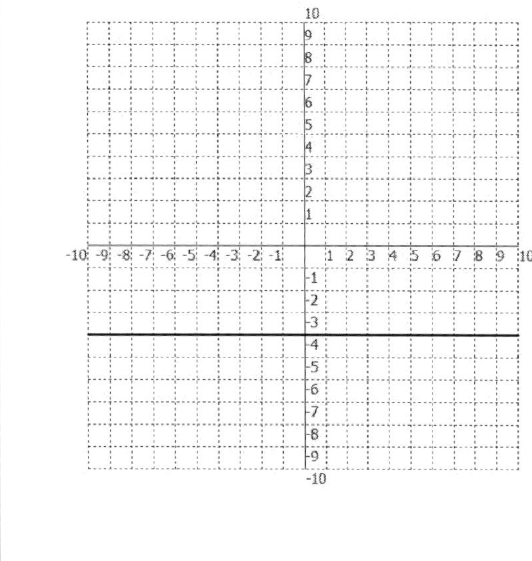	8. $y-4=-1$ $y=3$ 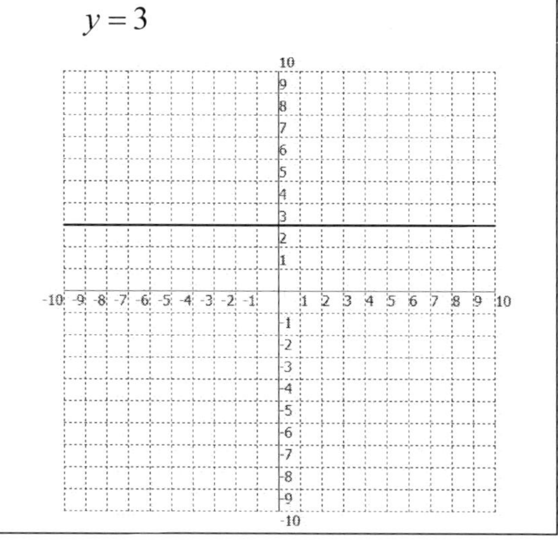

Finding Slope Given Two Points ~ Page 88

1. $m=\frac{4}{3}$	2. $m=-\frac{2}{3}$
3. $m=-\frac{3}{3}=-1$	4. $m=\frac{5}{5}=1$
5. $m=\frac{13-3}{5-1}=\frac{10}{4}=\frac{5}{2}$	6. $m=\frac{8-(-6)}{1-3}=\frac{14}{-2}=-7$
7. $m=\frac{-3-5}{0-4}=\frac{-8}{-4}=2$	8. $m=\frac{-2-(-2)}{2-(-4)}=\frac{0}{6}=0$

Finding Slope Given an Equation ~ Page 93

1.	2.
Slope is $\frac{2}{5}$.	$y-3x=1$ $y=3x+1$ Slope is 3.
3. $2y=5x+4$ $y=\frac{5}{2}x+2$ Slope is $\frac{5}{2}$.	4. $5y-10x=-15$ $5y=10x-15$ $y=2x-3$ Slope is 2.
5. $3x-2y=12$ $-2y=-3x+12$ $y=\frac{3}{2}x-6$ Slope is $\frac{3}{2}$.	6. $3x-4y-16=0$ $3x-4y=16$ $-4y=-3x+16$ $y=\frac{3}{4}x-4$ Slope is $\frac{3}{4}$.

Graphing a Linear Equation ~ Page 96

No Practice Problems

Equations of Parallel Lines ~ Page 98

1. $y=-2x+2$	2. $y=\frac{1}{2}x$
3. The first equation: $2y+2x=6$ $2y=-2x+6$ $y=-x+3$	4. $4x+6y=5$ $6y=-4x+5$ $y=-\frac{2}{3}x+\frac{5}{6}$ The first equation: $-3y=2x+5$ $y=-\frac{2}{3}x-\frac{5}{3}$

Writing a Linear Equation Given a Point and Slope ~ Page 100

1.	2.
$y = mx + b$ $4 = 2(1) + b$ $4 = 2 + b$ $2 = b$ $y = 2x + 2$	$y = mx + b$ $5 = 5(-6) + b$ $5 = -30 + b$ $35 = b$ $y = 5x + 35$
3.	**4.**
$y = mx + b$ $2 = \frac{1}{3}(-3) + b$ $2 = -1 + b$ $3 = b$ $y = \frac{1}{3}x + 3$	$y = mx + b$ $-3 = \frac{3}{4}(8) + b$ $-3 = 6 + b$ $-9 = b$ $y = \frac{3}{4}x - 9$

Writing a Linear Equation Given Two Points ~ Page 102

1.	2.
$m = \frac{6-2}{5-1} = \frac{4}{4} = 1$ $y = mx + b$ $2 = 1(1) + b$ $2 = 1 + b$ $1 = b$ $y = x + 1$	$m = \frac{4-(-1)}{3-2} = \frac{5}{1} = 5$ $y = mx + b$ $-1 = 5(2) + b$ $-1 = 10 + b$ $-11 = b$ $y = 5x - 11$
3.	**4.**
$m = \frac{-2-0}{3-(-3)} = \frac{-2}{6} = -\frac{1}{3}$ $y = mx + b$ $0 = -\frac{1}{3}(-3) + b$ $0 = 1 + b$ $-1 = b$ $y = -\frac{1}{3}x - 1$	$m = \frac{4-4}{2-(-2)} = \frac{0}{4} = 0$ $y = mx + b$ $4 = 0(-2) + b$ $4 = b$ $y = 4$

Graphing Inequalities ~ Page 104

1. $y<3$	2. $y>\frac{3}{2}x+2$
3.	4. 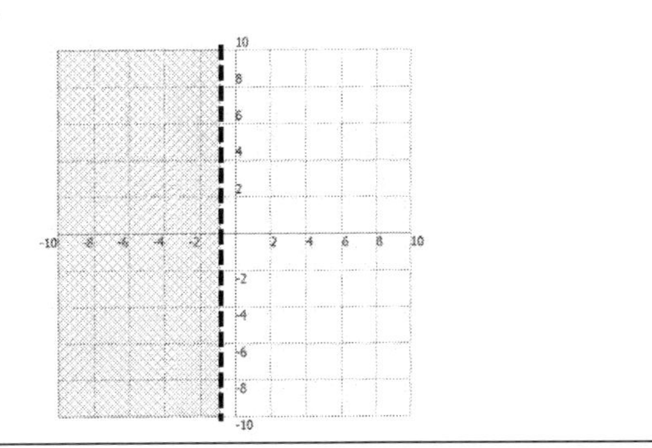
5. 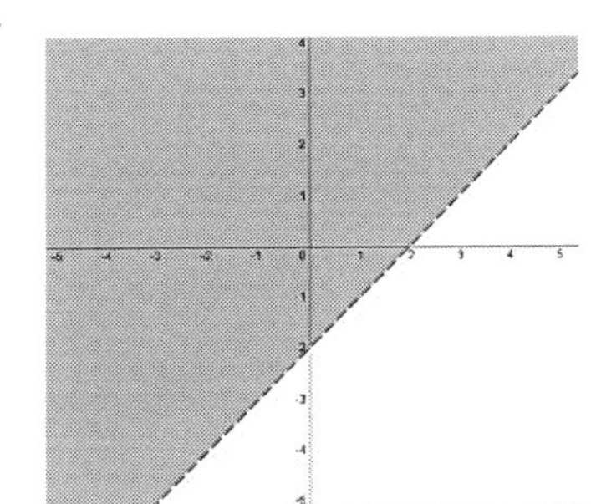	6. 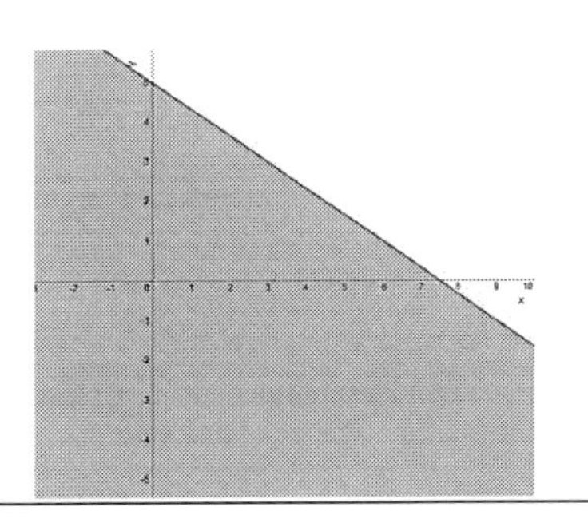
7. $x+y\le -3$ $y\le -x-3$ 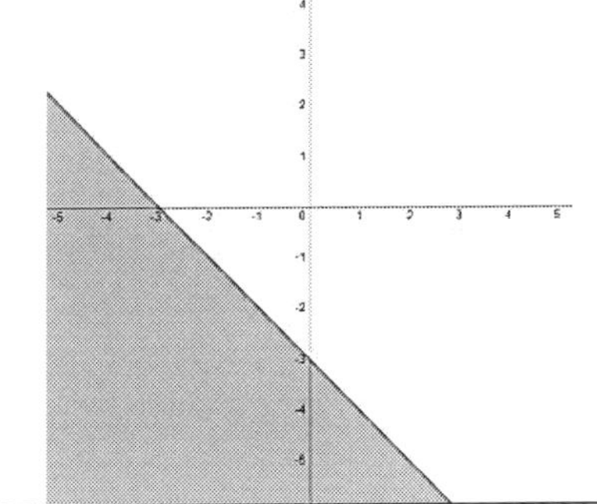	8. $x-y\le -1$ $-y\le -x-1$ $y\ge x+1$ 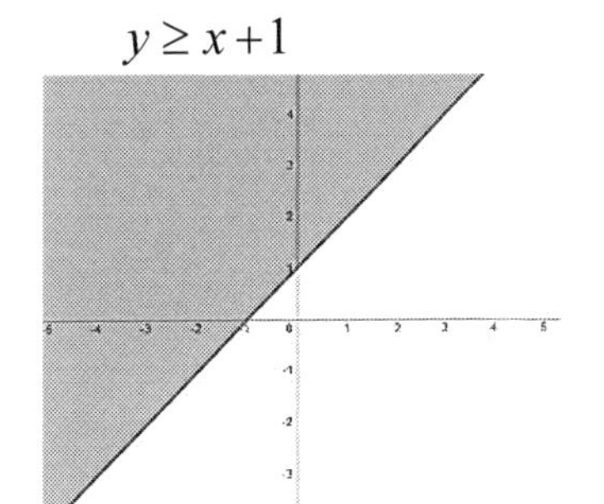
9. $2y-6x>10$ $2y>6x+10$ $y>3x+5$ 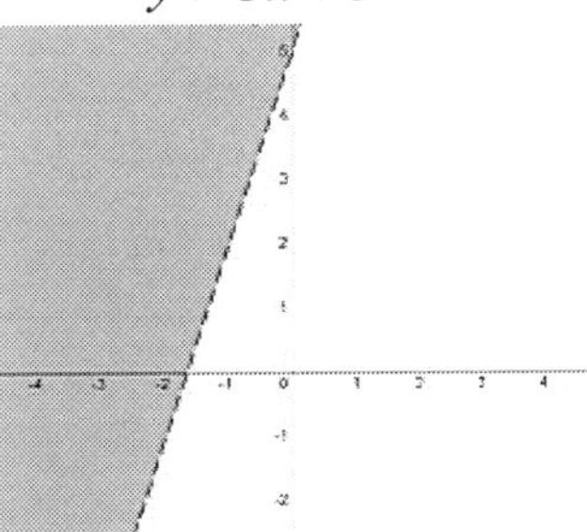	10. $9-x\ge 3y$ $3-\frac{1}{3}x\ge y$ $y\le -\frac{1}{3}x+3$

V. LINEAR SYSTEMS

Solving Systems of Equations Algebraically ~ Page 111

1. $\begin{array}{rr} 3x-y=8 & \\ \underline{x+y=4} & x+y=4 \\ 4x=12 & 3+y=4 \\ x=3 & y=1 \end{array}$	2. $\begin{array}{rr} 2x-3y=19 & 2(8)-3y=19 \\ \underline{3x+3y=21} & 16-3y=19 \\ 5x=40 & -3y=3 \\ x=8 & y=-1 \end{array}$
3. $\begin{array}{rr} 2x-4y=12 & \\ \underline{-2x+y=-9} & -2x-1=-9 \\ -3y=3 & -2x=-8 \\ y=-1 & x=4 \end{array}$	4. $\begin{array}{rr} 3x+y=0 & \\ \underline{-x-y=-4} & 3(-2)+y=0 \\ 2x=-4 & -6+y=0 \\ x=-2 & y=6 \end{array}$
5. $\begin{array}{rcr} 3x+2y=4 & \rightarrow & 3x+2y=4 \\ \underline{-2x+2y=24} & \underline{\times(-1)} & \underline{2x-2y=-24} \\ & & 5x=-20 \\ & & x=-4 \end{array}$ $\begin{array}{r} 3(-4)+2y=4 \\ -12+2y=4 \\ 2y=16 \\ y=8 \end{array}$	6. $\begin{array}{rcr} 2x+3y=6 & \rightarrow & 2x+3y=6 \\ \underline{2x+y=-2} & \underline{\times(-1)} & \underline{-2x-y=2} \\ & & 2y=8 \\ & & y=4 \end{array}$
7. $\begin{array}{rcr} -3x+4y=11 & \times 2 & -6x+8y=22 \\ \underline{6x-5y=-16} & \underline{\rightarrow} & \underline{6x-5y=-16} \\ & & 3y=6 \\ & & y=2 \end{array}$ $\begin{array}{r} -3x+4(2)=11 \\ -3x+8=11 \\ -3x=3 \\ x=-1 \end{array}$	8. $\begin{array}{rcr} 3x+4y=9 & \times 3 & 9x+12y=27 \\ \underline{5x+6y=21} & \underline{\times(-2)} & \underline{-10x-12y=-42} \\ & & -x=-15 \\ & & x=15 \end{array}$ $\begin{array}{r} 3(15)+4y=9 \\ 45+4y=9 \\ 4y=-36 \\ y=-9 \end{array}$
9. $\begin{array}{rr} 4x-10=5-x & \\ 5x-10=5 & \\ 5x=15 & y=5-3 \\ x=3 & y=2 \end{array}$	10. $\begin{array}{ll} x=(10-3x)-2 & y=10-3(2) \\ 4x=8 & y=4 \\ x=2 & \end{array}$

11.

$$3(9-2x)-2x=11$$
$$27-6x-2x=11$$
$$27-8x=11$$
$$-8x=-16$$
$$x=2$$

$$y=9-2(2)$$
$$y=5$$

12.

$$x-4y=-8$$
$$x=4y-8$$

$$7(4y-8)+3y=68$$
$$28y-56+3y=68$$
$$31y-56=68$$
$$31y=124$$
$$y=4$$

$$x-4(4)=-8$$
$$x-16=-8$$
$$x=8$$

Solving Systems of Equations Graphically ~ Page 118

1.

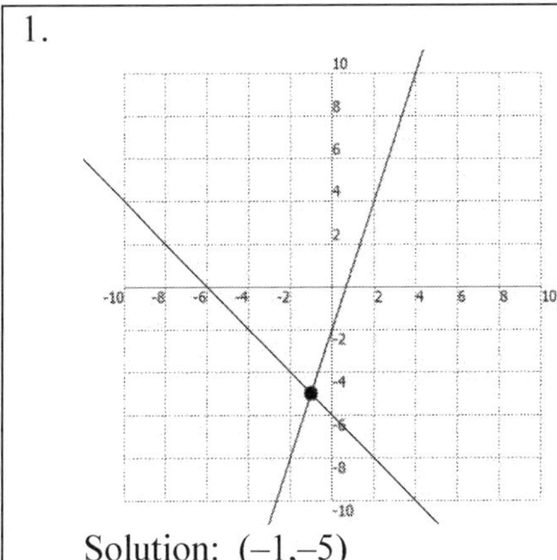

Solution: (–1,–5)

2.

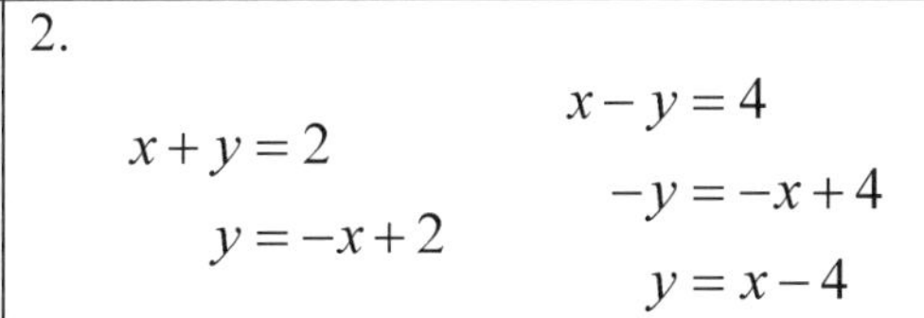

$$x+y=2$$
$$y=-x+2$$

$$x-y=4$$
$$-y=-x+4$$
$$y=x-4$$

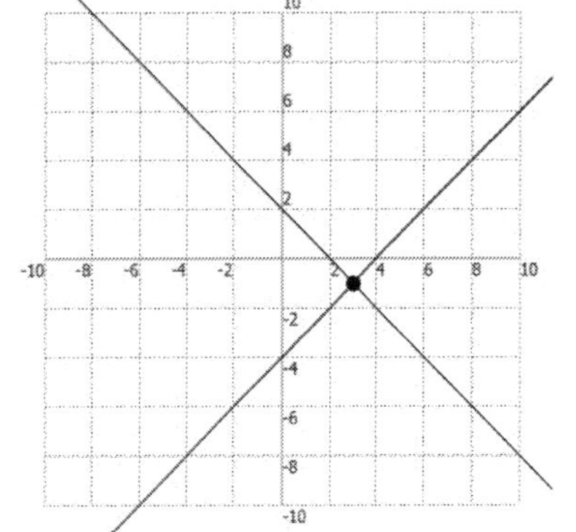

Solution: (3,–1)

3.

$$3x-5y=15$$
$$-5y=-3x+15$$
$$y=\tfrac{3}{5}x-3$$

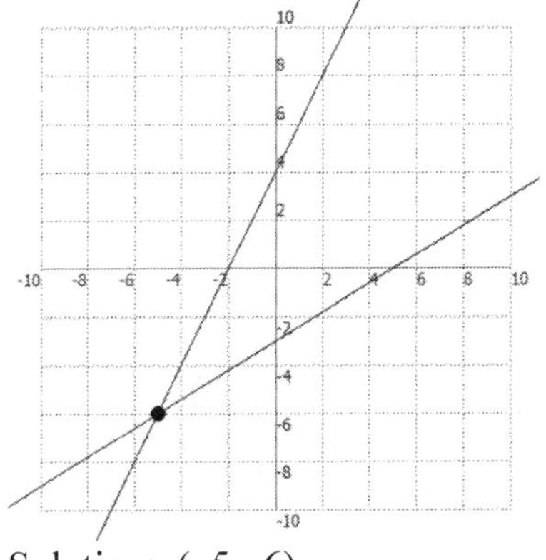

Solution: (–5,–6)

4.

$$x+3y=-3$$
$$3y=-x-3$$
$$y=-\tfrac{1}{3}x-1$$

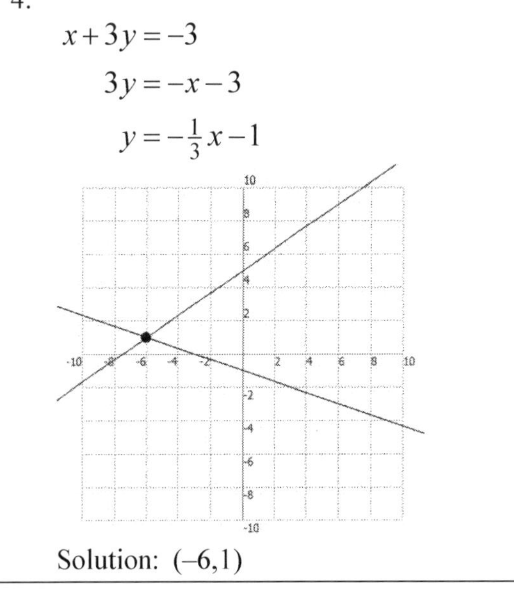

Solution: (–6,1)

Solving Systems of Inequalities Graphically ~ Page 122

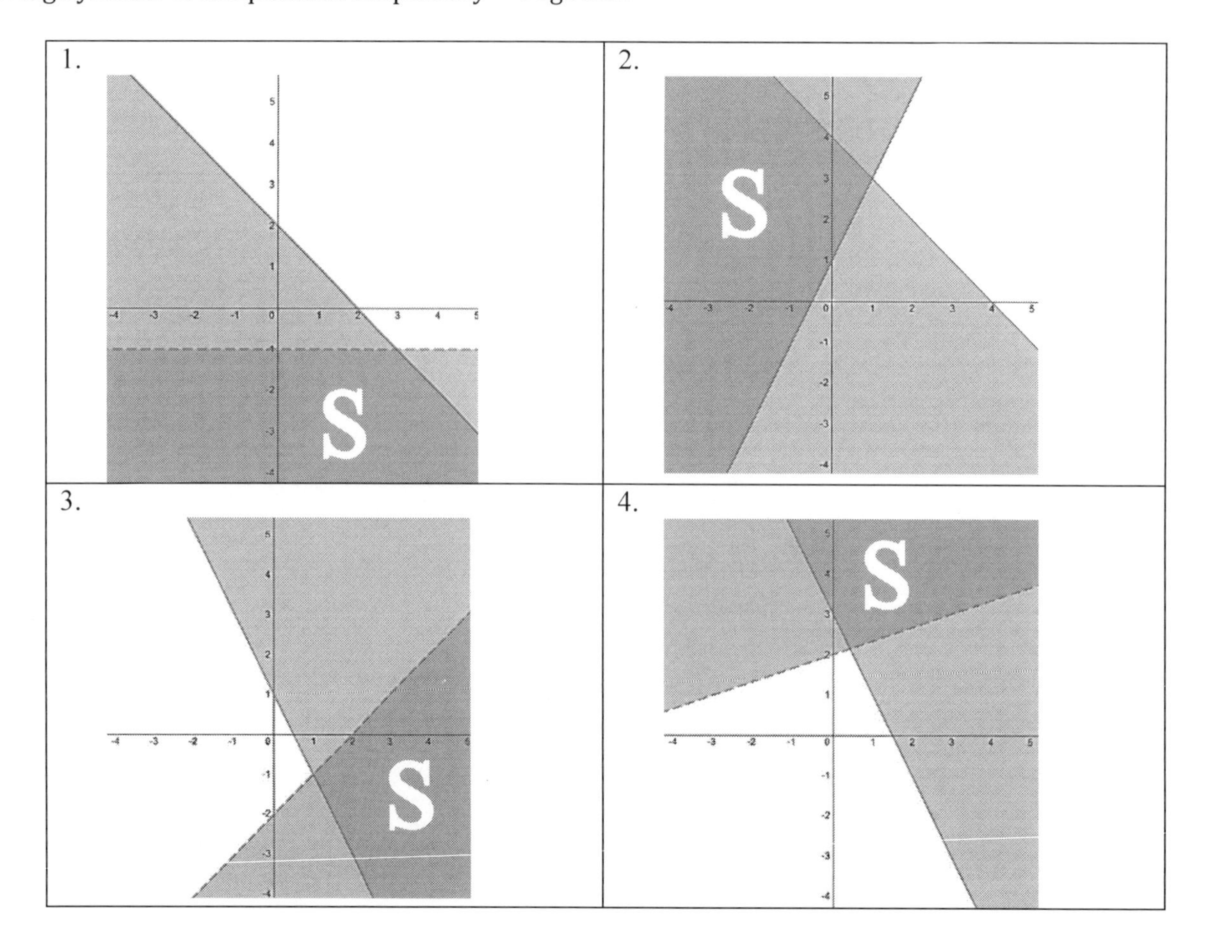

Solution Sets of Systems of Inequalities ~ Page 129

1. Yes	2. No
3. (a) (1,1)	4. (d) (–9,0)

Word Problems – Systems of Linear Equations ~ Page 133

1. x and y are the two numbers.

$$\begin{array}{r} x-y=5 \\ x+y=59 \\ \hline 2x=64 \\ x=32 \end{array} \qquad \begin{array}{r} (32)+y=59 \\ y=27 \end{array}$$

The numbers are 32 and 27.

2. d = cost of a doughnut
   c = cost of a cookie

$$\begin{array}{rl} 2d+3c=3.30 & \times 2 \\ 5d+2c=4.95 & \times(-3) \\ \hline \end{array}$$

$$\begin{array}{r} 4d+6c=6.60 \\ -15d-6c=-14.85 \\ \hline -11d=-8.25 \\ d=0.75 \end{array} \qquad \begin{array}{r} 2(0.75)+3c=3.30 \\ 1.50+3c=3.30 \\ 3c=1.80 \\ c=0.60 \end{array}$$

Doughnuts cost 75¢ and cookies cost 60¢.

3. p = cost of a pizza slice
   c = cost of a cola

$$\begin{array}{rl} 3p+2c=6.00 & \times 3 \\ 2p+3c=5.25 & \times(-2) \\ \hline \end{array}$$

$$\begin{array}{r} 9p+6c=18.00 \\ -4p-6c=-10.50 \\ \hline 5p=7.50 \\ p=1.50 \end{array} \qquad \begin{array}{r} 3(1.50)+2c=6.00 \\ 4.50+2c=6.00 \\ 2c=1.50 \\ c=0.75 \end{array}$$

Pizzas cost $1.50 and colas cost $0.75.

4. s = hourly rate for the sprayer
   g = hourly rate for the generator

$$\begin{array}{rl} 6s+6g=90 & \times 2 \\ 4s+8g=100 & \times(-3) \\ \hline \end{array}$$

$$\begin{array}{r} 12s+12g=180 \\ -12s-24g=-300 \\ \hline -12g=-120 \\ g=10 \end{array} \qquad \begin{array}{r} 6s+6(10)=90 \\ 6s+60=90 \\ 6s=30 \\ s=5 \end{array}$$

Sprayer costs $5/hr; generator costs $10/hr.

5. f = number of fancy shirts bought
   p = number of plain shirts bought

$$\begin{array}{rl} 28f+15p=131 & \rightarrow \\ f+p=7 & \times(-15) \\ \hline \end{array}$$

$$\begin{array}{r} 28f+15p=131 \\ -15f-15p=-105 \\ \hline 13f=26 \\ f=2 \end{array} \qquad \begin{array}{r} (2)+p=7 \\ p=5 \end{array}$$

She bought 2 fancy shirts and 5 plain shirts.

6. t = tens digit; u = units digit

$$\begin{array}{l} 10u+t=10t+u+9 \\ 9u+t=10t+9 \\ 9u-9t=9 \end{array} \qquad \begin{array}{rl} u+t=7 & \times 9 \\ 9u-9t=9 & \rightarrow \\ \hline \end{array}$$

$$\begin{array}{r} 9u+9t=63 \\ 9u-9t=9 \\ \hline 18u=72 \\ u=4 \end{array} \qquad \begin{array}{r} (4)+t=7 \\ t=3 \end{array}$$

The number is 34.

Word Problems – Systems of Inequalities ~ Page 137

<table>
<tr>
<td>1. d = number of dog-walking hours
c = number of car wash hours

$d + c \le 7$
$7.50d + 6.00c \ge 92.00$</td>
<td>2. s = number of bags of soil
p = number of plants

$4s + 10p \le 100$
$p \ge 5$</td>
</tr>
<tr>
<td>3. s = number of boxes of small books
l = number of boxes of large books

$15s + 8l \ge 350$
$s + l \ge 35$</td>
<td>4.
(a) $t \le 3$
$d \le 55t$
Therefore,
$d \le 55(3)$
$d \le 165$

(b) Yes</td>
</tr>
</table>

VI. POLYNOMIALS

Adding and Subtracting Polynomials ~ Page 139

1. $8x^2-1$	2. $4x^2+x-1$
3. $(3x^2+2xy+7)-(6x^2-4xy+3)=$ $3x^2+2xy+7-6x^2+4xy-3=$ $-3x^2+6xy+4$	4. $(a^2+a-1)-(3a^2-2a+5)=$ $a^2+a-1-3a^2+2a-5=$ $-2a^2+3a-6$
5. $(x^2-3x-2)-(2x^2-x+6)=$ $x^2-3x-2-2x^2+x-6=$ $-x^2-2x-8$	6. $(x^2+1)-(3x^2+4x-1)=$ $x^2+1-3x^2-4x+1=$ $-2x^2-4x+2$

Multiplying Polynomials ~ Page 144

1. $7x-7x^4$	2. $6r^3-15r$
3. $(c+8)(c-5)=$ $c^2-5c+8c-40=$ $c^2+3c-40$	4. $(x-7)(2x+3)=$ $2x^2+3x-14x-21=$ $2x^2-11x-21$
5. $a^2+2ab+b^2$	6. $(x-6)(x-6)=$ $x^2-6x-6x+36=$ $x^2-12x+26$
7. $\begin{array}{c\|c\|c} & x & 3 \\ \hline x & x^2 & 3x \\ \hline -y & -xy & -3y \\ \hline -1 & -x & -3 \end{array}$	8. (c) $ax+by$
9. $(x-1)(2x^2+x-2)=$ $2x^3+x^2-2x-2x^2-x+2=$ $2x^3-x^2-3x+2$	10. $(x^2+2)(x^2-2x+1)=$ $x^4-2x^3+x^2+2x^2-4x+2=$ $x^4-2x^3+3x^2-4x+2$

Dividing a Polynomial by a Monomial ~ Page 148

1. $\frac{2x+4}{2}=\frac{2x}{2}+\frac{4}{2}=x+2$	2. $\frac{x^2+2x}{x}=\frac{x^2}{x}+\frac{2x}{x}=x+2$
3. $\frac{14ab+28b}{14b}=\frac{14ab}{14b}+\frac{28b}{14b}=a+2$	4. $\frac{6x^3+9x^2+3x}{3x}=\frac{6x^3}{3x}+\frac{9x^2}{3x}+\frac{3x}{3x}=$ $2x^2+3x+1$

VII. RADICALS

Irrational Numbers ~ Page 150

1. (2) $\sqrt{8}$ $-\sqrt{16}=-4$, $\sqrt{64}=8$, $\sqrt{\frac{1}{64}}=\frac{1}{8}$	2. Irrational. 3 is not a perfect square.
3. Irrational. π is irrational, and the quotient of an irrational number and a non-zero rational number is irrational.	4. Irrational. $\sqrt{29}$ is irrational since 29 is not a perfect square. The numerator is the difference of a non-zero rational and irrational, so the numerator is irrational. The fraction is the quotient of an irrational and a non-zero rational, so it is irrational.
5. x could be 0, $\sqrt{3}$, $\sqrt{12}$, $\frac{1}{\sqrt{3}}$, etc.	6. $\pi \approx 3.141593$ and $\frac{22}{7} \approx 3.142857$ $\pi - 3.14 \approx 0.001593$ $\frac{22}{7} - \pi \approx 0.001264$ So, $\frac{22}{7}$ is a closer approximation.

Simplifying Radicals ~ Page 154

1. $\sqrt{12}=\sqrt{\boxed{2\cdot 2}\cdot 3}=2\sqrt{3}$	2. $\sqrt{50}=\sqrt{2\cdot\boxed{5\cdot 5}}=5\sqrt{2}$
3. $5\sqrt{72}=5\sqrt{\boxed{2\cdot 2}\cdot 2\cdot\boxed{3\cdot 3}}=$ $5\cdot 2\cdot 3\sqrt{2}=30\sqrt{2}$	4. $3\sqrt{45}=3\sqrt{\boxed{3\cdot 3}\cdot 5}=3\cdot 3\sqrt{5}=9\sqrt{5}$
5. $2\sqrt{128}=2\sqrt{\boxed{2\cdot 2}\cdot\boxed{2\cdot 2}\cdot\boxed{2\cdot 2}\cdot 2}=$ $2\cdot 2\cdot 2\cdot 2\sqrt{2}=16\sqrt{2}$	6. $\frac{7\sqrt{18}}{3}=\frac{7\cdot 3\sqrt{2}}{3}=7\sqrt{2}$

Operations with Radicals ~ Page 157

1. $\sqrt{75}+\sqrt{3}=5\sqrt{3}+\sqrt{3}=6\sqrt{3}$	2. $\sqrt{27}+\sqrt{12}=3\sqrt{3}+2\sqrt{3}=5\sqrt{3}$
3. $5\sqrt{7}+3\sqrt{28}=5\sqrt{7}+6\sqrt{7}=11\sqrt{7}$	4. $2\sqrt{50}-\sqrt{2}=10\sqrt{2}-\sqrt{2}=9\sqrt{2}$
5. $\sqrt{6}\cdot\sqrt{15}=\sqrt{90}=3\sqrt{10}$	6. $\sqrt{90}\cdot\sqrt{40}-\sqrt{8}\cdot\sqrt{18}=$ $\sqrt{3600}-\sqrt{144}=$ $60-12=48$
7. $\dfrac{\sqrt{65}}{\sqrt{5}}=\sqrt{13}$	8. $\dfrac{20\sqrt{100}}{4\sqrt{2}}=5\sqrt{50}=25\sqrt{2}$
9. $\dfrac{\sqrt{48}-5\sqrt{27}+2\sqrt{75}}{\sqrt{3}}=$ $\sqrt{16}-5\sqrt{9}+2\sqrt{25}=$ $4-15+10=-1$	10. $\dfrac{\sqrt{27}+\sqrt{75}}{\sqrt{12}}=\dfrac{3\sqrt{3}+5\sqrt{3}}{2\sqrt{3}}=\dfrac{8\sqrt{3}}{2\sqrt{3}}=4$

VIII. CLASSIFICATION OF STATISTICAL DATA

Population and Sample ~ Page 162

1. The population is all the bolts in the shipment. The sample is the 100 selected bolts.	2. The population is all the mall shoppers. The sample is every sixth person within the 3 hour period.

Qualitative and Quantitative Data ~ Page 163

1. Quantitative	2. Qualitative

Univariate and Bivariate Data ~ Page 166

1. Bivariate (temperature and revenue)	2. Univariate. One qualitative variable (pitch type) is needed to track each pitch.

IX. UNIVARIATE DATA

Dot Plots and Distributions ~ Page 170

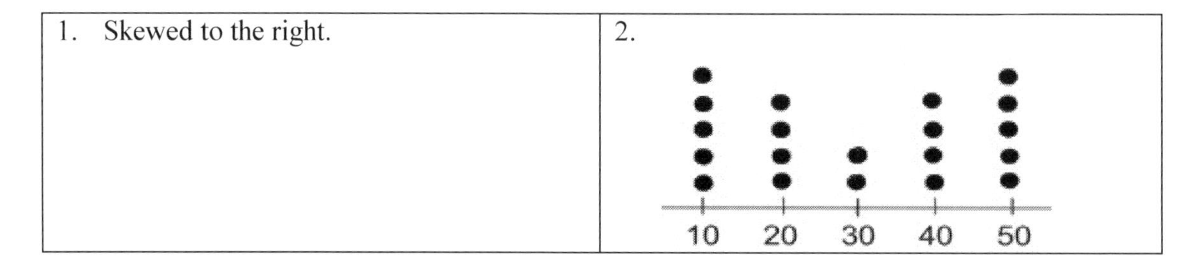

1. Skewed to the right.	2.

Frequency Tables and Histograms ~ Page 172

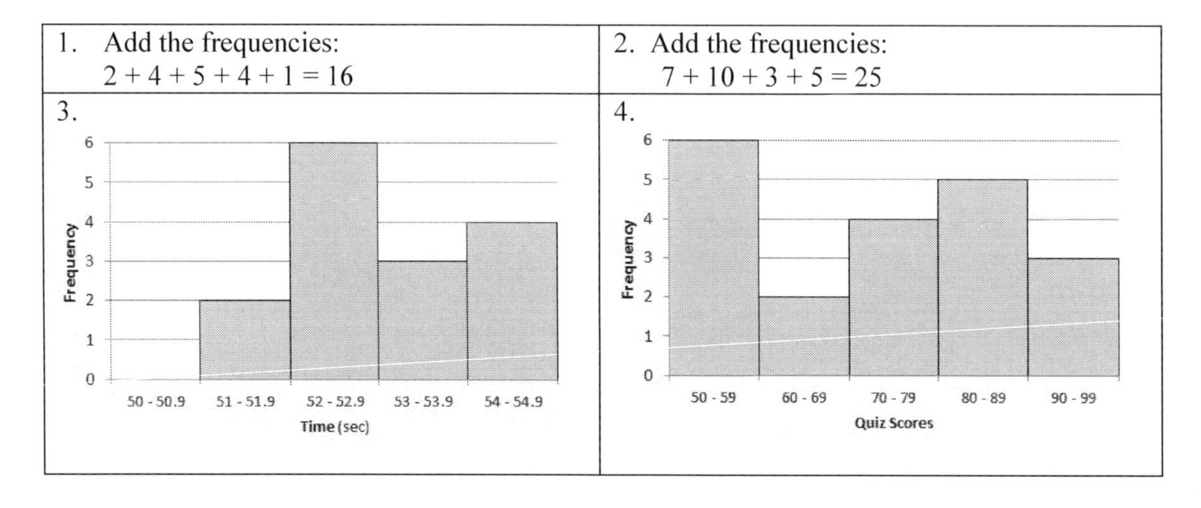

1. Add the frequencies: $2 + 4 + 5 + 4 + 1 = 16$	2. Add the frequencies: $7 + 10 + 3 + 5 = 25$
3.	4.

Central Tendency ~ Page 181

1. mode	2. median
3. (a)	4. They are all divided by two as well.
5. City A (22)	6. mean = 79, median = 79, mode = 78
7. mean = 22, median = 20, mode = 20	8. 131 – 150 There are 44 total scores, so the median would be the average of the 22nd and 23rd highest scores.

Standard Deviation ~ Page 189

1. mean = 66, sample SD ≈ 30.4	2. mean ≈ 60.7, sample SD ≈ 15.1
3. mean = 44, sample SD ≈ 4.0	4. mean = 7.2, sample SD ≈ 3.7
5. mean = $610, sample SD ≈ 14.7	6. The first set, as shown by the smaller SD.

Percentiles ~ Page 194

1. $\frac{22}{30} = 73\frac{1}{3}\%$, so the 73rd percentile.	2. 359. 40%, or 240 out of the 600, have GPAs below Tony's. So, Tony is ranked 360th highest (600 – 240), which means that 359 students have higher GPAs.
3. $p = \frac{b+h}{n} = \frac{5+0.5}{11} = 0.5$, so 70 is the 50th percentile.	4. $p = \frac{b+h}{n} = \frac{18+1}{25} = 0.76$, So 90 is the 76th percentile.

Quartiles ~ Page 197

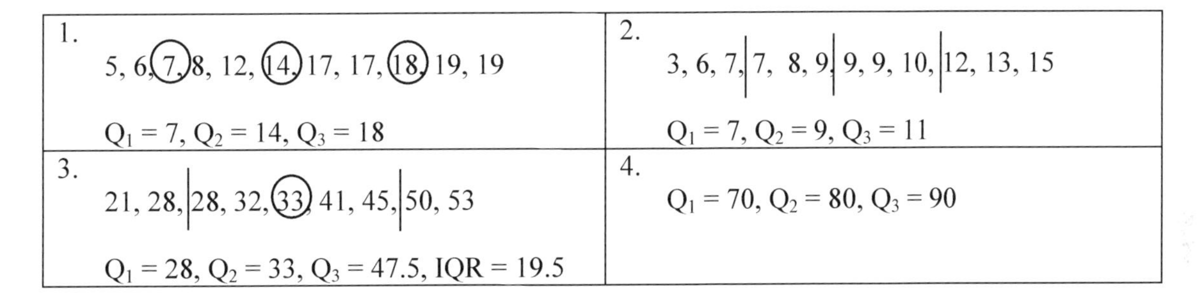

1. 5, 6, (7), 8, 12, (14), 17, 17, (18), 19, 19 $Q_1 = 7$, $Q_2 = 14$, $Q_3 = 18$	2. 3, 6, 7, | 7, 8, 9 | 9, 9, 10, | 12, 13, 15 $Q_1 = 7$, $Q_2 = 9$, $Q_3 = 11$
3. 21, 28, | 28, 32, (33), 41, 45, | 50, 53 $Q_1 = 28$, $Q_2 = 33$, $Q_3 = 47.5$, IQR = 19.5	4. $Q_1 = 70$, $Q_2 = 80$, $Q_3 = 90$

Box Plots ~ Page 201

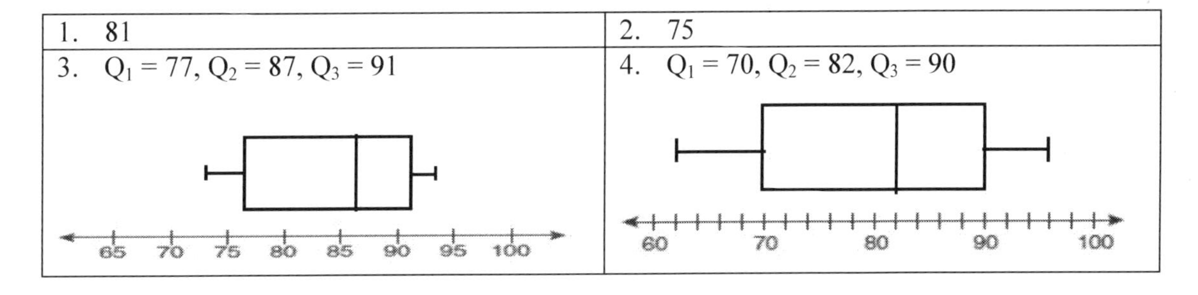

1. 81	2. 75
3. $Q_1 = 77$, $Q_2 = 87$, $Q_3 = 91$ 65 70 75 80 85 90 95 100	4. $Q_1 = 70$, $Q_2 = 82$, $Q_3 = 90$ 60 70 80 90 100

X. BIVARIATE DATA

Two-Way Frequency Tables ~ Page 210

1.

$\frac{15}{113} \approx 13.3\%$ of the students are undecided.

$\frac{31}{60} \approx 51.7\%$ of the 9th graders are watching.

2.

	Fiction	Nonfiction	Total
Hardcover	28	52	80
Paperback	94	36	130
Total	122	88	210

	Fiction	Nonfiction	Total
Hardcover	13.3%	24.8%	38.1%
Paperback	44.8%	17.1%	61.9%
Total	58.1%	41.9%	100%

3. Given data in bold below.

	Coca-Cola	Sprite	Total
Table	16	**14**	30
Garbage	34	8	**42**
Total	**50**	22	**72**

Scatter Plots ~ Page 213

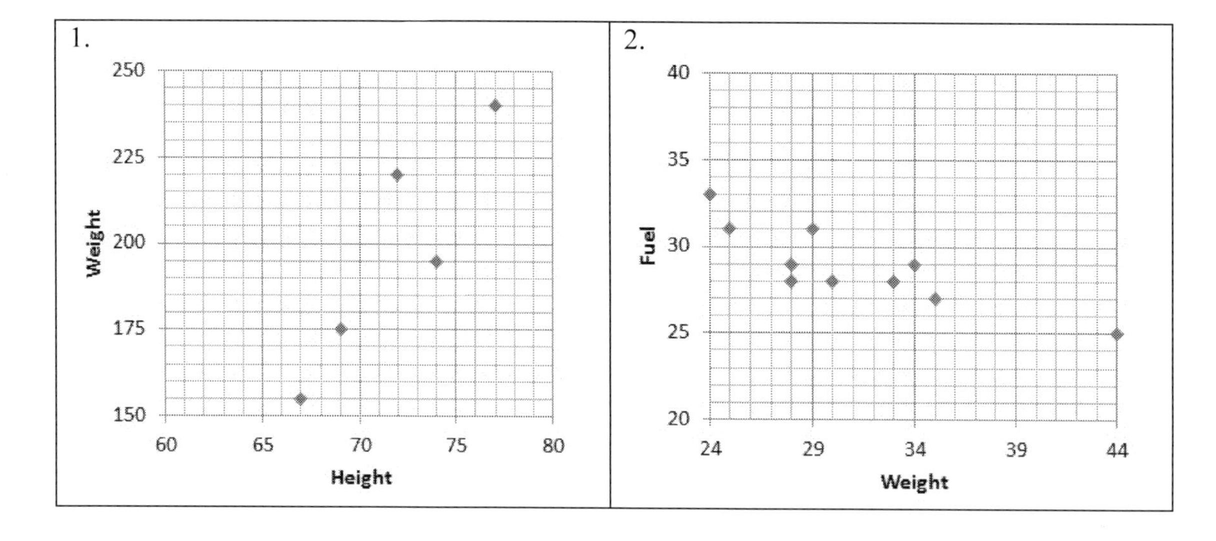

Correlation and Causality ~ Page 220

1.	2.
(a) positive: children usually gain weight as they age and grow (b) negative: as the volume of water increases, the remaining space decreases (c) none: shoe size and hair length are unrelated (d) positive: more people go to the beach when the temperature is higher	A. positive, causal B. negative, causal C. positive, not causal, the size and severity of the fire, which results in more firefighters being called D. negative, not causal, the degree of civilization and industrialization over time

Identifying Correlation in Scatter Plots ~ Page 224

1. (a)	2. strong positive
3. weak negative	

Lines of Fit ~ Page 230

1. (a) 80 wpm (b) 9 wpm	2. Line A. Most of the points are closer to Line A than to Line B.
3. $700, $600, $500, $400, $300, $200, $100, $0 10 12 14 16 18 20 22 24 26 Temperature °C	4. $y = 0.2x + 7.5$
5. $y = 2x + 5.14$	6. (c) $y = 1,000x + 15,000$
7. The prediction for the 35 year old is more likely to be accurate, since it is an interpolation rather than an extrapolation.	8. 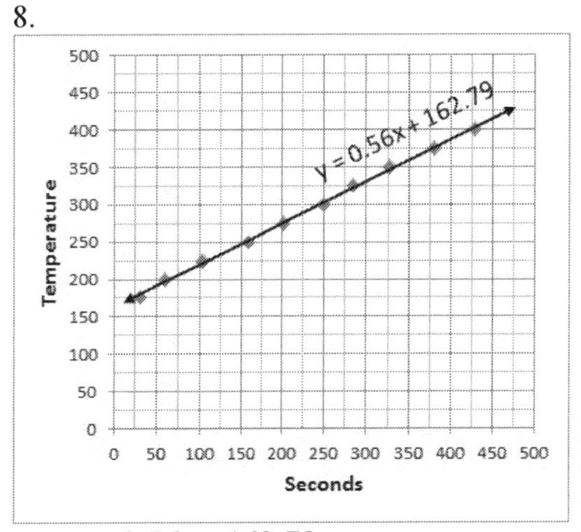 $y = 0.56x + 162.79$

Residuals and Correlation Coefficients ~ Page 241

1. Actual value – Predicted value = 12,550 – 14,050 = –1,500

2. (b) –0.24
It is a weak correlation.

3.
a. 0.90 b. –0.40 c. 0.99
d. –0.85 e. 0.50 f. 0

4.
$r = 0.9858534782$

5.

Study Time in Hours (x)	Test Score (y)	Predicted Test Score	Residual
0.5	63	62.8	0.2
1	67	67.2	-0.2
1.5	72	71.6	0.4
2	76	76.0	0
2.5	80	80.4	-0.4
3	85	84.8	0.2
3.5	89	89.2	-0.2

6.
(a) $y = 0.75(22) - 0.25 = 16.25$.
(Since football scores cannot be fractional, 16 is a valid answer.)
(b) $y = 0.75(34) - 0.25 = 25.25$
Residual = 32 – 25.25 = 6.75
(c) $y = 0.75(28) - 0.25 = 20.75$
$p - 20.75 = -0.75$
$p = 20$
They scored 20 points.

Residual Plots ~ Page 248

1.

x	y	Predicted Value	Residual
5	3	2.5	0.5
10	4	5.0	-1
15	9	7.5	1.5
20	7	10.0	-3
25	13	12.5	0.5
30	15	15.0	0

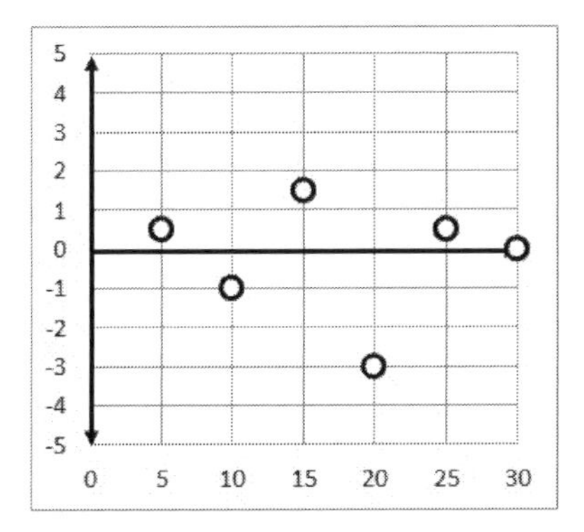

Yes. There is no clear pattern in the residual plot.

2.

x	y	Predicted Value	Residual
2	5	15.5	-10.5
4	15	14.7	0.3
6	26	13.9	12.1
8	23	13.1	9.9
10	11	12.3	-1.3
12	3	11.5	-8.5

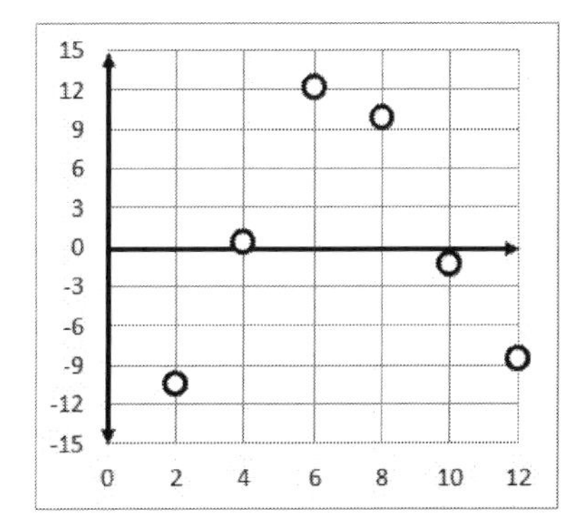

No. There appears to be a parabola-like pattern in the residual plot.

3.

(a) $y = 0.117x + 83.267$

(b) and (c)

Distance (miles)	Airfare ($)	Predicted Price ($)	Residual
576	178	150.7	27.3
370	138	126.6	11.4
612	94	154.9	-60.9
1,216	278	225.5	52.5
409	158	131.1	26.9
1,502	258	259.0	-1.0
946	198	193.9	4.1
998	188	200.0	-12.0
189	98	105.4	-7.4
787	179	175.3	3.7
210	138	107.8	30.2
737	98	169.5	-71.5

(d)

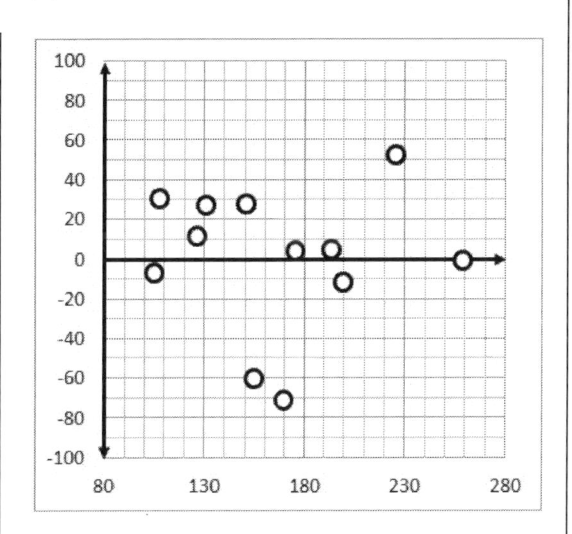

XI. INTRODUCTION TO FUNCTIONS

Determining if Relations are Functions ~ Page 252

1. (b)	2. (b)
3. Yes, each x entry is mapped to a unique y	

Determining if Graphs Represent Functions ~ Page 254

1. (a)	2. (b)
3. (d)	4. (a)

Function Notation, Domain and Range ~ Page 260

1. $\{2, 3, 22, 51\}$	2. $x \neq 0$
3. (a) $f(x) \geq 0$ (b) $0 \leq f(x) \leq 9$	4. the set of counting (natural) numbers
5. (a) $f(n) = 5n$ (b) whole numbers $n \leq 20$ (c) $\{0, 5, 10, 15, \ldots 100\}$	

Function Graphs ~ Page 263
No Practice Problems

Evaluating Functions ~ Page 266

1. $f(3) = -2(3)^2 - 3(3) - 6 =$ $-18 - 9 - 6 = -33$	2. $f(2) = 0.5^2 = 0.25$
3. $g(4a) = 2(4a)^2 + 6(4a) - 3 =$ $32a^2 + 24a - 3$	4. $h(0) = 2(0) - 1 = -1$ $h(-2) = 2(-2) - 1 = -5$ $h(0) \cdot h(-2) = (-1)(-5) = 5$
5. $f(a+2) = (a+2)^2 + 2(a+2) - 1 =$ $a^2 + 4a + 4 + 2a + 4 - 1 =$ $a^2 + 6a + 7$	6. $-10 = -4x + 2$ $-12 = -4x$ $3 = x$

Operations on Functions ~ Page 269

1. $h(x)=(x^2+x+1)+(x-5)=$ x^2+2x-4	2. $h(x)=(2x+1)(x-2)=$ $2x^2-3x-2$
3. (a) $R(c)=20c+500$ (b) $E(c)=6c$ (c) $P(c)=R(c)-E(c)=$ $(20c+500)-(6c)=14c+500$	

Rate of Change for Linear Functions ~ Page 270

1. negative	2. positive; $m=\dfrac{348-232}{6-4}=\dfrac{116}{2}=58$ mph

Average Rate of Change ~ Page 274

1. $f(5)=5^2+2=27$ $f(15)=15^2+2=227$ $m=\dfrac{227-27}{15-5}=\dfrac{200}{10}=20$	2. $f(-3)=(-3)^2+10(-3)+16=-5$ $f(3)=3^2+10(3)+16=55$ $m=\dfrac{55-(-5)}{3-(-3)}=\dfrac{60}{6}=10$

Word Problems – Function Graphs ~ Page 278

1. 7 minutes
   From 7:04 to 7:07 and 7:20 to 7:24

2.
   (a) Spencer starts at (0,20) and McKenna starts at (0,0).
   (b) McKenna speeds up, as the graph curves upward. The average rate of change increases.
   (c) At about 3.2 hours. They traveled about 41 miles.

3.

Characteristic of Graph	Interpretation in Terms of the Race
y-intercepts	At 11 A.M. Runner A is 10 miles from the finish line and Runner B is 7 miles from the finish line.
Slopes	Runner A is decreasing the distance to the finish line at 8 mph and Runner B is running at 3.5 mph.
Point of intersection	The two runners meet after about 2/3 hour at about $4\frac{2}{3}$ miles from the finish line.
x-intercepts	Runner A finishes at 12:15 and Runner B finishes at 1:00.

XII. EXPONENTIAL FUNCTIONS

Percent Change ~ Page 285

1. $\frac{\lvert 86-95\rvert}{86} \approx 10.5\%$	2. $\frac{\lvert 160-116\rvert}{160} = 27.5\%$
3. $\frac{\lvert 10-12\rvert}{10} = 20\%$	4. $\frac{\lvert 4.2-6\rvert}{4.2} \approx 42.9\%$
5. $\frac{\lvert 20-12\rvert}{20} = 40\%$	6. $\frac{\lvert 40-34\rvert}{40} = 15\%$

Exponential Growth and Decay ~ Page 288

1. $x(1.1)^{20}$	2. $x(0.98)^{n}$
3. $\$1500(1.05)^{6} \approx \$2010.14$	4. $25{,}000(0.8)^{3} = 12{,}800$
5. $3{,}810(1.035)^{5} \approx 4{,}525$	6. $256(0.25)^{3} = 4$
7. $\$11{,}900(0.87)^{3} \approx \$7{,}800$	8. $\$1.39(1.005)^{12} \approx \$1.48$

Exponential Functions ~ Page 293

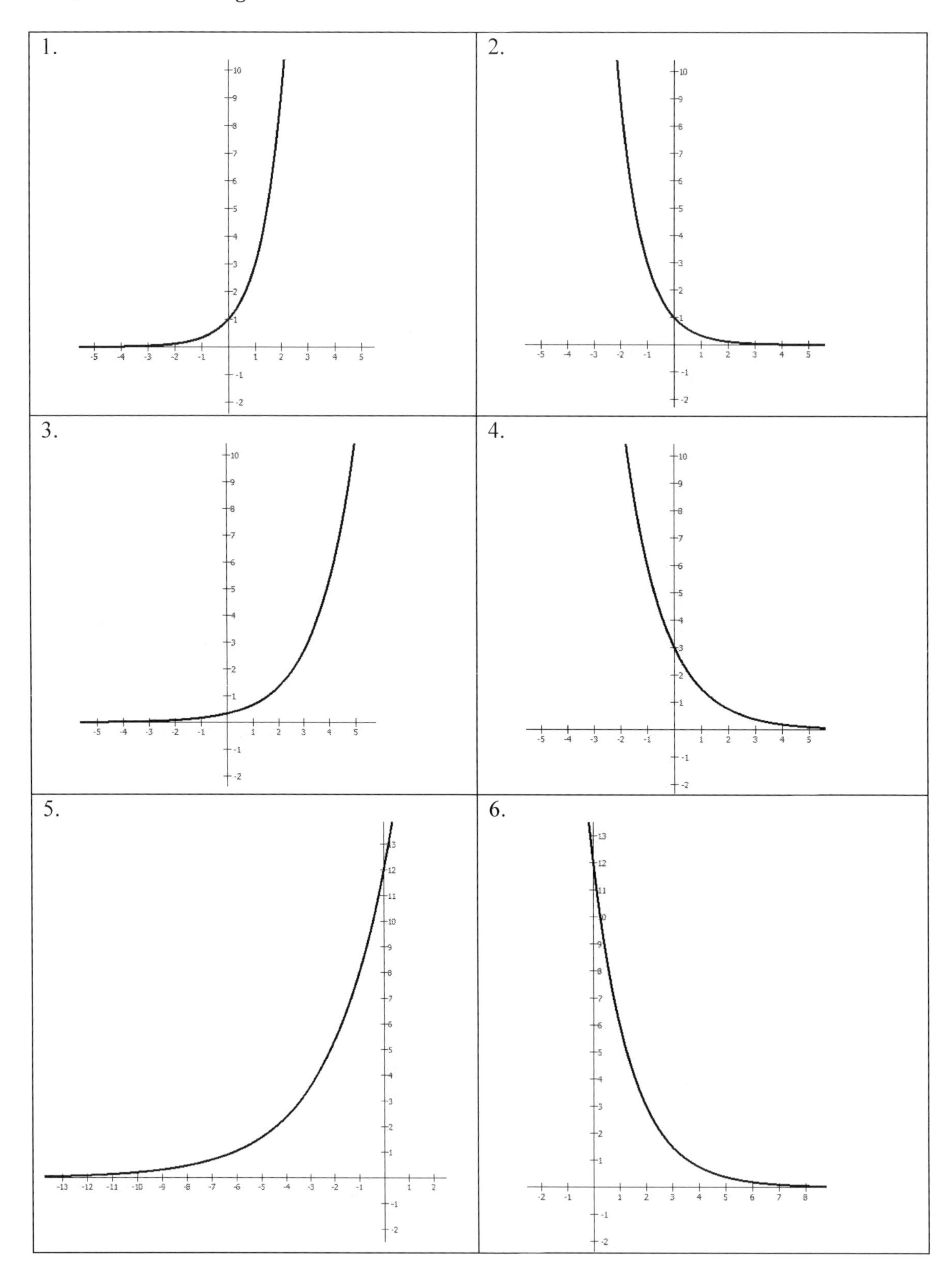

7.

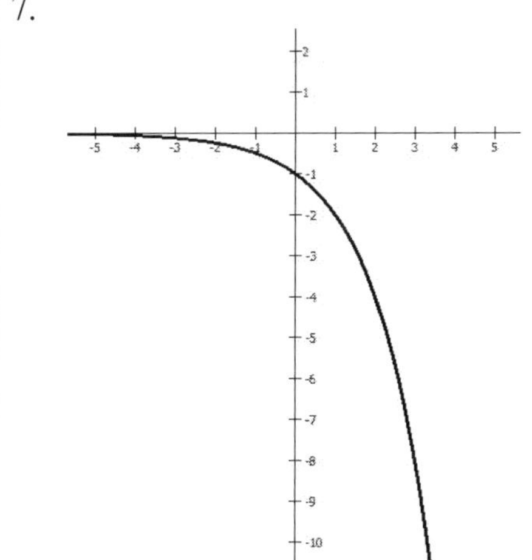

8.

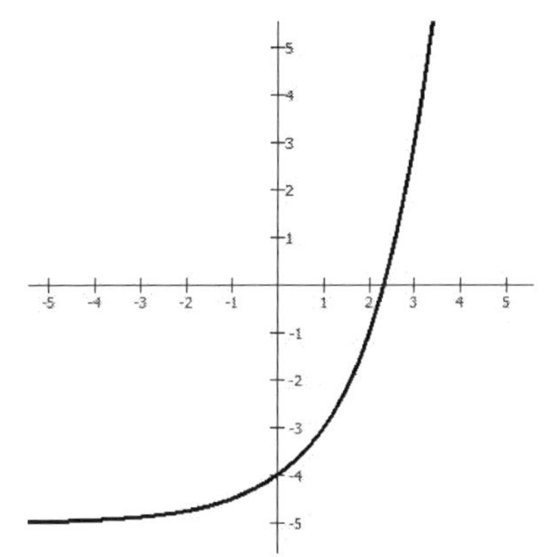

It is shifted (translated) down by 5 units.

9.

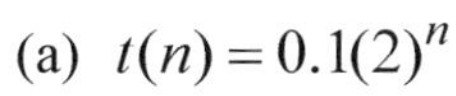

(a) $t(n) = 0.1(2)^n$

(b)

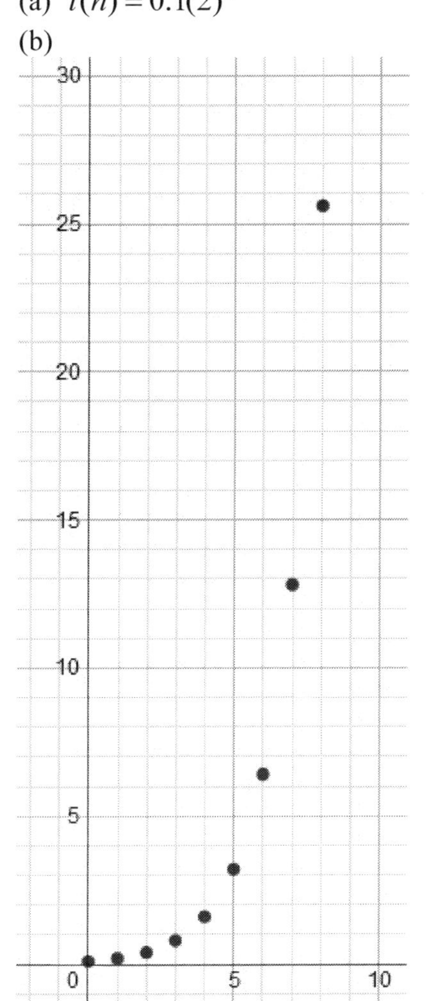

10.

(a) $h(n) = 30\left(\frac{1}{2}\right)^n$

(b)

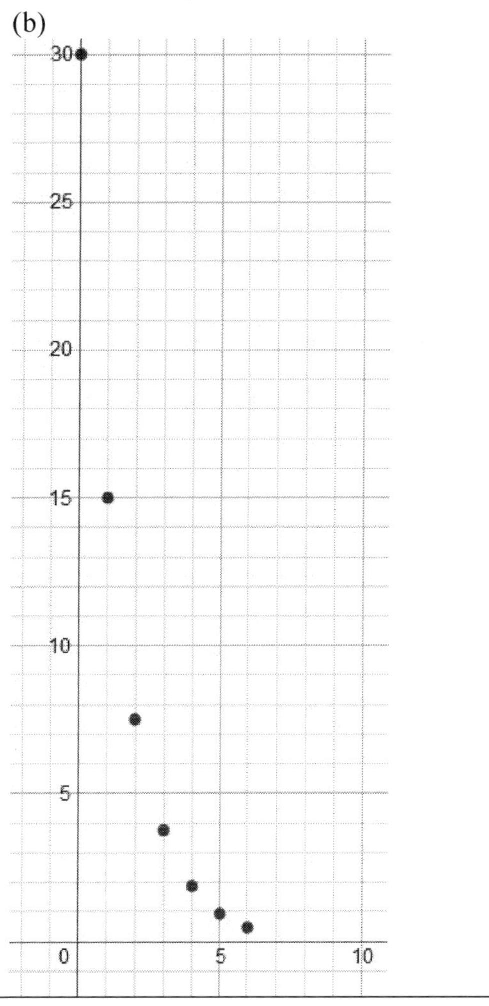

Sequences ~ Page 299

1. –4	2. $\frac{1}{2}$
3. $a_n = a_1 + (n-1)d$ $a_n = 15 + (n-1) \cdot 5$ $a_n = 15 + 5n - 5$ $a_n = 5n + 10$	4. $a_n = a_1 r^{n-1}$ $a_n = (-1)(-2)^{n-1}$ $a_n = -(-2)^{n-1}$
5. $a_n = a_1 + (n-1)d$ $a_n = 21 + (8-1) \cdot 9$ $a_n = 84$	6. $a_n = a_1 r^{n-1}$ $a_n = 6\left(-\frac{1}{2}\right)^{7-1}$ $a_n = \frac{6}{64} = 0.09375$
7. 3, 5, 8, 12 neither; there is no common difference nor common ratio	8. (a) $\begin{cases} f(1) = 40 \\ f(2) = 8 \\ f(n) = \dfrac{f(n-2) + f(n-1)}{2} \text{ for } n > 2 \end{cases}$ (b) $f(7) = 19$

Comparing Linear and Exponential Functions ~ Page 304

1. (a) linear (c) exponential (b) exponential (d) neither	2. $f(n) = 3n$
3. $f(n) = 3^n$	

XIII. FACTORING

Factoring Out the Greatest Common Factor ~ Page 307

1. $4x^2-6x=2\cdot2\cdot x\cdot x-2\cdot3\cdot x$ $=2x(2x-3)$	2. $5a^2-10a=5\cdot a\cdot a-2\cdot5\cdot a$ $=5a(a-2)$
3. $7x(2x^2+1)$	4. $x(x^2+x-1)$
5. $6xy(2x^2+3y)$	6. $2y(y^2-2y+1)$
7. $3x(x^2-2x+2)$	8. $-2(x+y)$

Factoring a Trinomial ~ Page 310

1. $(x+7)(x+2)$	2. $(x-9)(x-2)$
3. $(x-9)(x+3)$	4. $(a-15)(a+14)$
5. yes, it is prime	6. $(-3x^2+x-2)+(4x^2+3x-10)=$ $x^2+4x-12=(x+6)(x-2)$

Factoring the Difference of Two Perfect Squares ~ Page 313

1. $(x+6)(x-6)$	2. $(2x+3)(2x-3)$
3. $(3+x)(3-x)$	4. $(a+1)(a-1)$
5. $(7x+y)(7x-y)$	6. $(2a+3b)(2a-3b)$
7. $(xy+4)(xy-4)$	8. $(x^5+10)(x^5-10)$

Factoring Completely ~ Page 316

1.	2.
$2y^2+12y-54=$ $2(y^2+6y-27)=$ $2(y+9)(y-3)$	$3x^2+15x-42=$ $3(x^2+5x-14)=$ $3(x+7)(x-2)$
3.	4.
$3x^2-27=$ $3(x^2-9)=$ $3(x+3)(x-3)$	$2x^2-50=$ $2(x^2-25)=$ $2(x+5)(x-5)$
5.	6.
$2a^2-10a-28=$ $2(a^2-5a-14)=$ $2(a-7)(a+2)$	$x^3+8x^2+7x=$ $x(x^2+8x+7)=$ $x(x+7)(x+1)$
7.	8.
$2x^8+16x^7+32x^6=$ $2x^6(x^2+8x+16)=$ $2x^6(x+4)(x+4)$	$3ax^2-27a=$ $3a(x^2-9)=$ $3a(x+3)(x-3)$
9.	10.
$5x^2y^3-180y=$ $5y(x^2y^2-36)=$ $5y(xy+6)(xy-6)$	$2x^5-32x=$ $2x(x^4-16)=$ $2x(x^2+4)(x^2-4)=$ $2x(x^2+4)(x+2)(x-2)$

Factoring by Grouping ~ Page 320

No Practice Problems

Factoring Trinomials with Leads Other than 1 ~ Page 322

1.	2.
$6x^2+x-2=$ $6x^2-3x+4x-2=$ $3x(2x-1)+2(2x-1)=$ $(3x+2)(2x-1)$	$12x^2+5x-2=$ $12x^2-3x+8x-2=$ $3x(4x-1)+2(4x-1)=$ $(3x+2)(4x-1)$
3. $12x^2-29x+15=$ $12x^2-9x-20x+15=$ $3x(4x-3)-5(4x-3)=$ $(3x-5)(4x-3)$	4. $6x^2-11x+4=$ $6x^2-3x-8x+4=$ $3x(2x-1)-4(2x-1)=$ $(3x-4)(2x-1)$
5. $15x^2+14x-8=$ $15x^2+20x-6x-8=$ $5x(3x+4)-2(3x+4)=$ $(5x-2)(3x+4)$	6. $-10x^2-29x-10=$ $-10x^2-25x-4x-10=$ $-5x(2x+5)-2(2x+5)=$ $(-5x-2)(2x+5)$
7. $4x^2+12x+9=$ $4x^2+6x+6x+9=$ $2x(2x+3)+3(2x+3)=$ $(2x+3)(2x+3)$ The square root is $2x+3$.	8. $10x^2+11x-6=$ $10x^2+15x-4x-6=$ $5x(2x+3)-2(2x+3)=$ $(5x-2)(2x+3)$
9. $5x^2-50x+120=$ $5(x^2-10x+24)=$ $5(x-6)(x-4)$	10. $12x^3+14x^2-6x=$ $2x[6x^2+7x-3]=$ $2x[6x^2-2x+9x-3]=$ $2x[2x(3x-1)+3(3x-1)]=$ $2x(2x+3)(3x-1)$

XIV. QUADRATIC EQUATIONS

Solving Quadratic Equations ~ Page 325

1. $x^2-5x=0$ $x(x-5)=0$ $x=0 \text{ or } x-5=0$ $x=0 \text{ or } x=5$	2. $x^2+3x-18=0$ $(x+6)(x-3)=0$ $x+6=0 \text{ or } x-3=0$ $x=-6 \text{ or } x=3$
3. $4x^2-36=0$ $4(x^2-9)=0$ $4(x+3)(x-3)=0$ $x+3=0 \text{ or } x-3=0$ $x=-3 \text{ or } x=3$	4. $x^2-5x-24=0$ $(x-8)(x+3)=0$ $x-8=0 \text{ or } x+3=0$ $x=8 \text{ or } x=-3$
5. $x^2-5x=6$ $x^2-5x-6=0$ $(x-6)(x+1)=0$ $x-6=0 \text{ or } x+1=0$ $x=6 \text{ or } x=-1$	6. $x^2-3=2x$ $x^2-2x-3=0$ $(x-3)(x+1)=0$ $x-3=0 \text{ or } x+1=0$ $x=3 \text{ or } x=-1$
7. $x^2-4x=x+24$ $x^2-5x-24=0$ $(x-8)(x+3)=0$ $x-8=0 \text{ or } x+3=0$ $x=8 \text{ or } x=-3$	8. $2x^2+10x=12$ $2x^2+10x-12=0$ $2(x^2+5x-6)=0$ $2(x+6)(x-1)=0$ $x+6=0 \text{ or } x-1=0$ $x=-6 \text{ or } x=1$
9. $x(x+2)=3$ $x^2+2x=3$ $x^2+2x-3=0$ $(x+3)(x-1)=0$ $x+3=0 \text{ or } x-1=0$ $x=-3 \text{ or } x=1$	10. $(x+2)(x+3)=12$ $x^2+5x+6=12$ $x^2+5x-6=0$ $(x+6)(x-1)=0$ $x+6=0 \text{ or } x-1=0$ $x=-6 \text{ or } x=1$

Finding Quadratic Equations from Given Roots ~ Page 330

1.	2.
$x=10$ *or* $x=-2$ $x-10=0$ *or* $x+2=0$ $(x-10)(x+2)=0$ $x^2-8x-20=0$	$x=0$ *or* $x=3$ $x=0$ *or* $x-3=0$ $x(x-3)=0$ $x^2-3x=0$
3. multiply both sides by 2 $x-\frac{3}{2}=0$ *or* $x-2=0$ $2x-3=0$ *or* $x-2=0$ $(2x-3)(x-2)=0$ $2x^2-7x+6=0$	4. $(x-1)^2=0$ $x^2-2x+1=0$
5. $(x-4)(x+4)=0$ $x^2-16=0$	6. $x=0$ *or* $x=1$ *or* $x=-1$ $x(x-1)(x+1)=0$ $x(x^2-1)=0$ $x^3-x=0$

Undefined Expressions ~ Page 332

1.	2.
$x+2=0$ $x=-2$	$3x+1=0$ $3x=-1$ $x=-\frac{1}{3}$
3. $x^2-4=0$ $(x+2)(x-2)=0$ $x=-2$ *or* $x=2$	4. $x^2+2x-8=0$ $(x+4)(x-2)=0$ $x=-4$ *or* $x=2$

Solving Proportions by Quadratic Equations ~ Page 335

1.	2.
$x(x+1)=6$ $x^2+x-6=0$ $(x+3)(x-2)=0$ $\{-3,2\}$	$5(x+3)=x(x+7)$ $5x+15=x^2+7x$ $x^2+2x-15=0$ $(x+5)(x-3)=0$ $\{-5,3\}$
3.	4.
$(x+1)(x-1)=8$ $x^2-1=8$ $x^2-9=0$ $(x+3)(x-3)=0$ $\{-3,3\}$	$2x(x-1)=x(3+x)$ $2x^2-2x=3x+x^2$ $x^2-5x=0$ $x(x-5)=0$ $\{0,5\}$

Completing the Square ~ Page 337

1.	2.
$\frac{b^2}{4a}=\frac{100}{4}=25$ $x^2+10x-11=0$ $x^2+10x=11$ $x^2+10x+25=11+25$ $(x+5)^2=36$ $x+5=\sqrt{36}$ $x+5=\pm 6$ $x=-5\pm 6$ Solution: $\{1,-11\}$	$\frac{b^2}{4a}=\frac{64}{4}=16$ $x^2-8x+16=0$ $x^2-8x=-16$ $x^2-8x+16=-16+16$ $(x-4)^2=0$ $x-4=0$ Solution: $\{4\}$
3.	4.
$\frac{b^2}{4a}=\frac{16}{4}=4$ $x^2+4x+2=0$ ← *Typo in text!* $x^2+4x=-2$ $x^2+4x+4=-2+4$ $(x+2)^2=2$ $x+2=\pm\sqrt{2}$ $x=-2\pm\sqrt{2}$ Solution: $\{-2+\sqrt{2},-2-\sqrt{2}\}$	$\frac{b^2}{4a}=\frac{16}{4}=4$ $x^2-4x-8=0$ $x^2-4x=8$ $x^2-4x+4=8+4$ $(x-2)^2=12$ $x-2=\pm\sqrt{12}=\pm 2\sqrt{3}$ $x=2\pm 2\sqrt{3}$ Solution: $\{2+2\sqrt{3},2-2\sqrt{3}\}$
5.	6.
$\frac{b^2}{4a}=\frac{4}{4}=1$ $x^2-2x+3=0$ $x^2-2x=-3$ $x^2-2x+1=-3+1$ $(x-1)^2=-2$ $x-1=\boxed{\pm\sqrt{-2}}$ No real solutions, since the square root of a negative number is not a real number.	$w(w+10)=880$ $\frac{b^2}{4a}=\frac{100}{4}=25$ $w^2+10w=880$ $w^2+10w+25=880+25$ $(w+5)^2=905$ $w+5=\pm\sqrt{905}$ $w=-5\pm\sqrt{905}\approx -5\pm 30.1$ $w=25.1$ (*reject negative w*) width ≈ 25.1 ft. and length ≈ 35.1 ft.

7. a is a perfect square.

$$\frac{b^2}{4a}=\frac{16}{16}=1$$

$$4x^2-4x-5=0$$
$$4x^2-4x=5$$
$$4x^2-4x+1=5+1$$
$$(2x-1)^2=6$$
$$2x-1=\pm\sqrt{6}$$
$$2x=1\pm\sqrt{6}$$
$$x=\frac{1\pm\sqrt{6}}{2}$$

Solution: $\left\{\frac{1+\sqrt{6}}{2},\frac{1-\sqrt{6}}{2}\right\}$

8. a is a perfect square.

$$\frac{b^2}{4a}=\frac{64}{16}=4$$

$$4x^2+8x=45$$
$$4x^2+8x+4=45+4$$
$$(2x+2)^2=49$$
$$2x+2=\pm\sqrt{49}=\pm7$$
$$2x=-2\pm7$$
$$x=\frac{-2\pm7}{2}$$

Solution: $\left\{\frac{5}{2},-\frac{9}{2}\right\}$

9.

$$3x^2-2x-1=0$$
$$9x^2-6x-3=0$$

$$\frac{b^2}{4a}=\frac{36}{36}=1$$

$$9x^2-6x-3=0$$
$$9x^2-6x=3$$
$$9x^2-6x+1=3+1$$
$$(3x-1)^2=4$$
$$3x+1=\pm\sqrt{4}=\pm2$$
$$3x=-1\pm2$$
$$x=\frac{-1\pm2}{3}$$

Solution: $\left\{\frac{1}{3},-1\right\}$

10.

$$5x^2+10x-25=0$$
$$x^2+2x-5=0$$

$$\frac{b^2}{4a}=\frac{4}{4}=1$$

$$x^2+2x-5=0$$
$$x^2+2x=5$$
$$x^2+2x+1=5+1$$
$$(x+1)^2=6$$
$$x+1=\pm\sqrt{6}$$
$$x=-1\pm\sqrt{6}$$

Solution: $\left\{-1+\sqrt{6},-1-\sqrt{6}\right\}$

Quadratic Formula and the Discriminant ~ Page 342

1. (c) 2	2. $b^2 - 4ac = -12$, so the answer is (a) imaginary
3. $x^2 + 7x + 8 = 0$ $\frac{-b \pm \sqrt{b^2 - 4ac}}{2a} =$ $\frac{-(7) \pm \sqrt{(7)^2 - 4(1)(8)}}{2(1)} =$ $\frac{-7 \pm \sqrt{17}}{2}$ Solution: $\left\{\frac{-7+\sqrt{17}}{2}, \frac{-7-\sqrt{17}}{2}\right\}$	4. $2x^2 - 8x + 3 = 0$ $\frac{-b \pm \sqrt{b^2 - 4ac}}{2a} =$ $\frac{-(-8) \pm \sqrt{(-8)^2 - 4(2)(3)}}{2(2)} =$ $\frac{8 \pm \sqrt{40}}{4} = \frac{8 \pm 2\sqrt{10}}{4} =$ $2 \pm \frac{\sqrt{10}}{2}$ Solution: $\left\{2 + \frac{\sqrt{10}}{2}, 2 - \frac{\sqrt{10}}{2}\right\}$

Word Problems – Quadratic Equations ~ Page 346

1.	2.
$x^2-2x=48$ $x^2-2x-48=0$ $(x-8)(x+6)=0$ $x-8=0$ *or* $x+6=0$ $x=8$ (*reject* $x=-6$) The number is 8.	$x^2+(x+8)^2=104$ $x^2+x^2+16x+64=104$ $2x^2+16x-40=0$ $2(x^2+8x-20)=0$ $2(x+10)(x-2)=0$ $x=2$ (*reject* $x=-10$) Numbers are 2 and 10.
3.	4.
$w(w+5)=500$ $w^2+5w-500=0$ $(w+25)(w-20)=0$ $w=20$ (*reject* $w=-25$) Width is 20 and length is 25.	$(t+7)(t-3)=24$ $t^2+4t-21=24$ $t^2+4t-45=0$ $(t+9)(t-5)=0$ $t=5$ (*reject* $t=-9$) Tamara is 5 years old.
5.	6.
$x(x+2)=63$ $x^2+2x-63=0$ $(x+9)(x-7)=0$ $x=-9$ (*reject* $x=7$) Numbers are –9 and –7.	$x(x+2)=(x+4)+8$ $x^2+2x=x+12$ $x^2+x-12=0$ $(x+4)(x-3)=0$ $x=3$ (*reject* $x=-4$) Numbers are 3, 5, and 7.
7.	8.
$(x+1)(x+10)=90$ $x^2+11x+10=90$ $x^2+11x-80=0$ $(x+16)(x-5)=0$ $x=5$ (*reject* $x=-16$) Numbers are 5 and 6.	$x(x+4)=2(x+2)+20$ $x^2+4x=2x+4+20$ $x^2+2x-24=0$ $(x+6)(x-4)=0$ $x=4$ (*reject* $x=-6$) Ages are 4, 6, and 8

XV. PARABOLAS

Finding Roots Given a Parabolic Graph ~ Page 351

1. 2 and 4	2. zero (only)
3. –2 and 3	4. –4 and 2

Finding Vertex and Axis of Symmetry Graphically ~ Page 356

1. Vertex is (3,–1); axis of symmetry is $x = 3$.	2. Vertex is (0,0); axis of symmetry is $x = 0$.
3. Vertex is (–1,7); axis of symmetry is $x = -1$.	4. Vertex is (–2,–3); axis of symmetry is $x = -2$.

Finding Vertex and Axis of Symmetry Algebraically ~ Page 363

1. $x = \frac{-b}{2a} = \frac{-4}{-2} = 2$ $y = -x^2 + 4x - 8$ $y = -(2)^2 + 4(2) - 8 = -4$ Vertex is (2,–4); axis of symmetry is $x = 2$.	2. $x = \frac{-b}{2a} = \frac{6}{2} = 3$ $y = x^2 - 6x + 10$ $y = (3)^2 - 6(3) + 10 = 1$ Vertex is (3,1); axis of symmetry is $x = 3$.
3. $x = \frac{-b}{2a} = \frac{-6}{6} = -1$ $y = 3(-1)^2 + 6(-1) - 1 = -4$ Vertex is (–1,–4).	4. $x = \frac{-b}{2a} = \frac{-8}{4} = -2$ $y = 2(-2)^2 + 8(-2) + 9 = 1$ Vertex (minimum) is (–2,1).
5. $x = \frac{-b}{2a} = \frac{-2}{2} = -1$ $y = (-1)^2 + 2(-1) = -1$ Vertex is (–1,–1); axis of symmetry is $x = -1$.	6. $x = \frac{-b}{2a} = \frac{0}{6} = 0$ $y = 3(0)^2 + 1 = 1$ Vertex is (0,1); axis of symmetry is $x = 0$.

Graphing Parabolas ~ Page 367

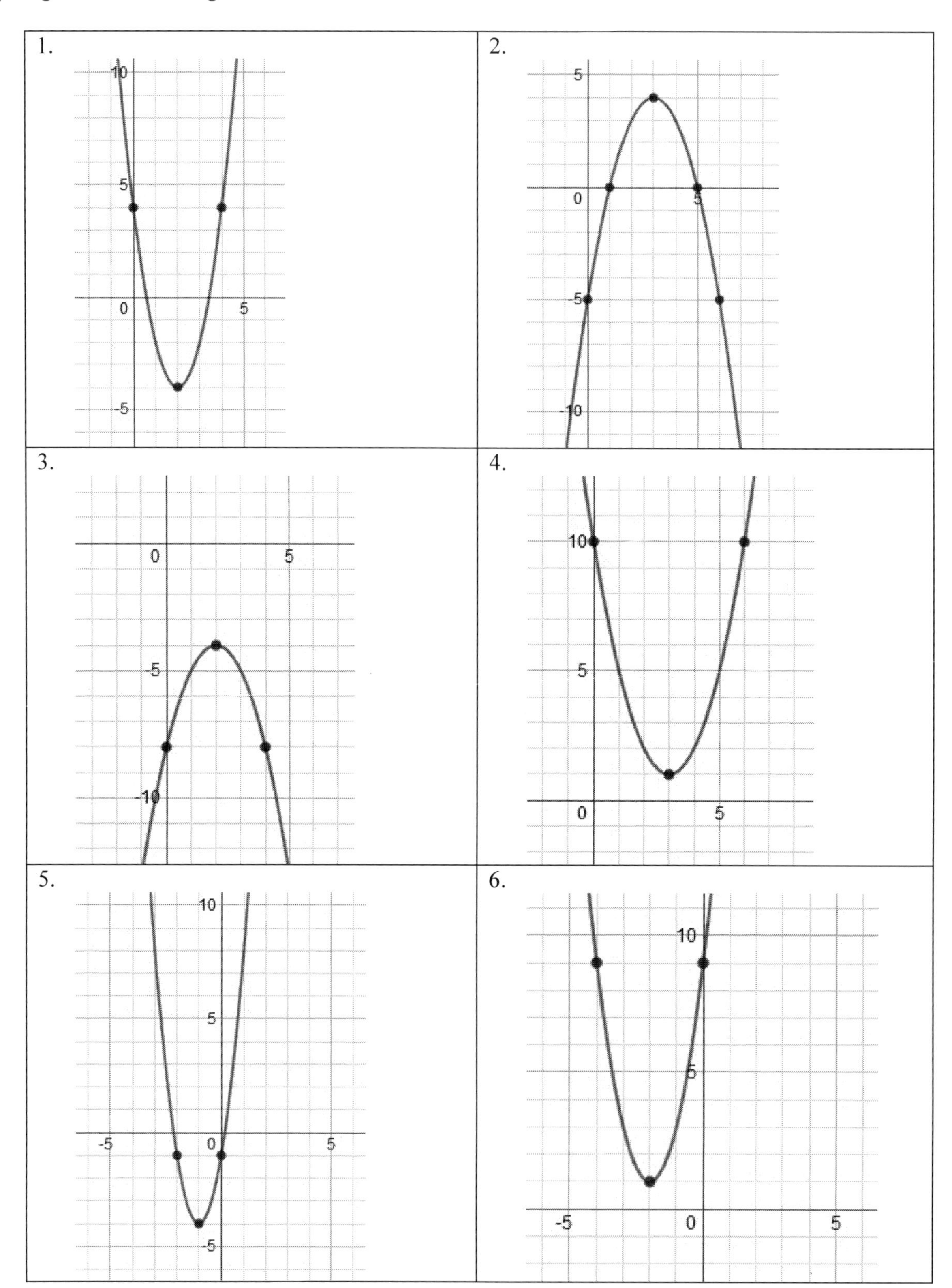

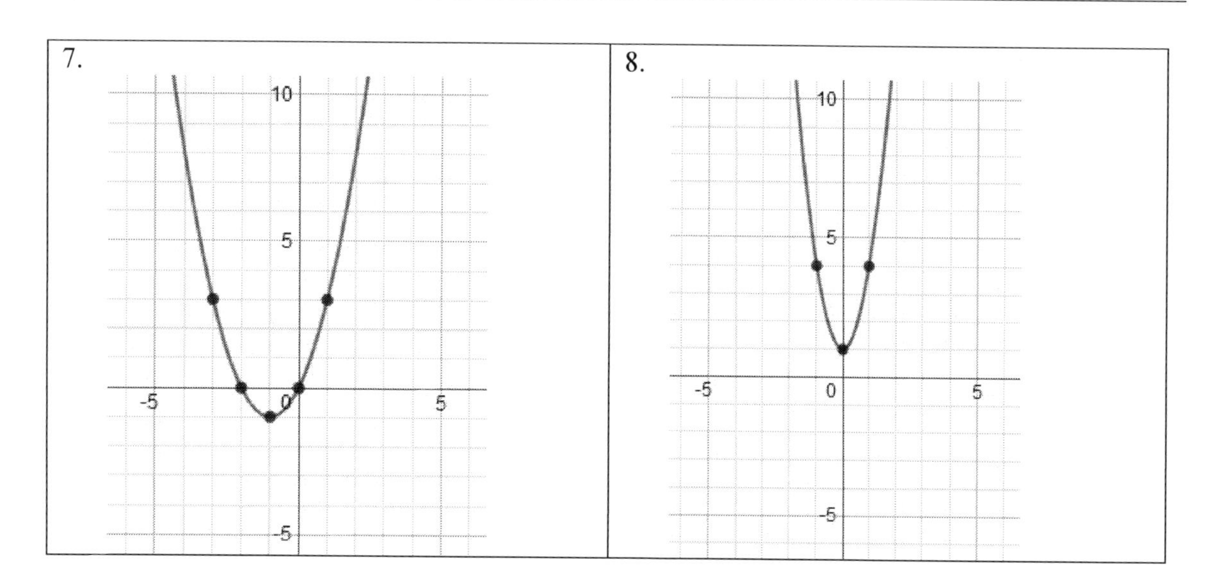
7.
10
5
-5
0
5
-5
8.
10
5
-5
0
5
-5

XVI. QUADRATIC-LINEAR SYSTEMS

Solving Quadratic-Linear Systems Algebraically ~ Page 372

1.	2.
$x^2-5=-4x$ $x^2+4x-5=0$ $(x+5)(x-1)=0$ $x=\{-5,1\}$ When $x=-5, y=-4(-5)=20$ When $x=1, y=-4(1)=-4$ Solutions: (–5,20) and (1,–4)	$x^2+4x+1=5x+3$ $x^2-x-2=0$ $(x-2)(x+1)=0$ $x=\{2,-1\}$ When $x=2, y=5(2)+3=13$ When $x=-1, y=5(-1)+3=-2$ Solutions are (2,13) and (–1,–2)
3. $x^2+2x-1=3x+5$ $x^2-x-6=0$ $(x-3)(x+2)=0$ $x=\{3,-2\}$ When $x=3, y=3(3)+5=14$ When $x=-2, y=3(-2)+5=1$ Solutions: (3,14) and (–2,1)	4. $x^2+4x-2=2x+1$ $x^2+2x-3=0$ $(x+3)(x-1)=0$ $x=\{-3,1\}$ When $x=-3, y=2(-3)+1=-5$ When $x=1, y=2(1)+1=3$ Solutions: (–3,–5) and (1,3)
5. $y+3x=1 \rightarrow y=-3x+1$ $x^2+7x+22=-3x+1$ $x^2+10x+21=0$ $(x+7)(x+3)=0$ $x=\{-7,-3\}$ When $x=-7, y=-3(-7)+1=22$ When $x=-3, y=-3(-3)+1=10$ Solutions: (–7,22) and (–3,10)	6. $y+3x=6 \rightarrow y=-3x+6$ $x^2=y+2x+6 \rightarrow y=x^2-2x-6$ $x^2-2x-6=-3x+6$ $x^2+x-12=0$ $(x+4)(x-3)=0$ $x=\{-4,3\}$ When $x=-4, y=-3(-4)+6=18$ When $x=3, y=-3(3)+6=-3$ Solutions: (–4,18) and (3,–3)

Solving Quadratic-Linear Systems Graphically ~ Page 375

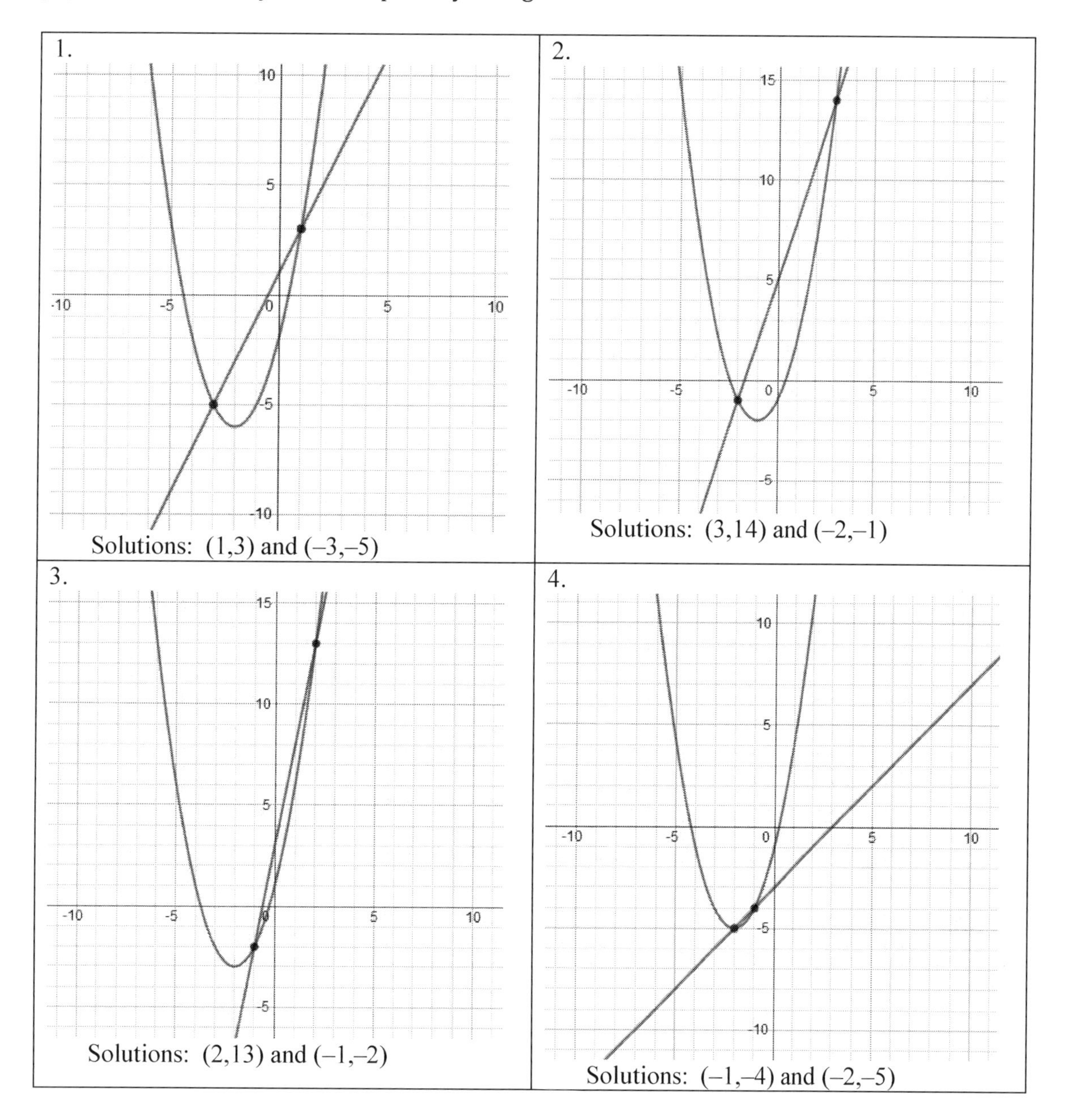

XVII. OTHER FUNCTIONS AND TRANSFORMATIONS

Absolute Value Functions ~ Page 384

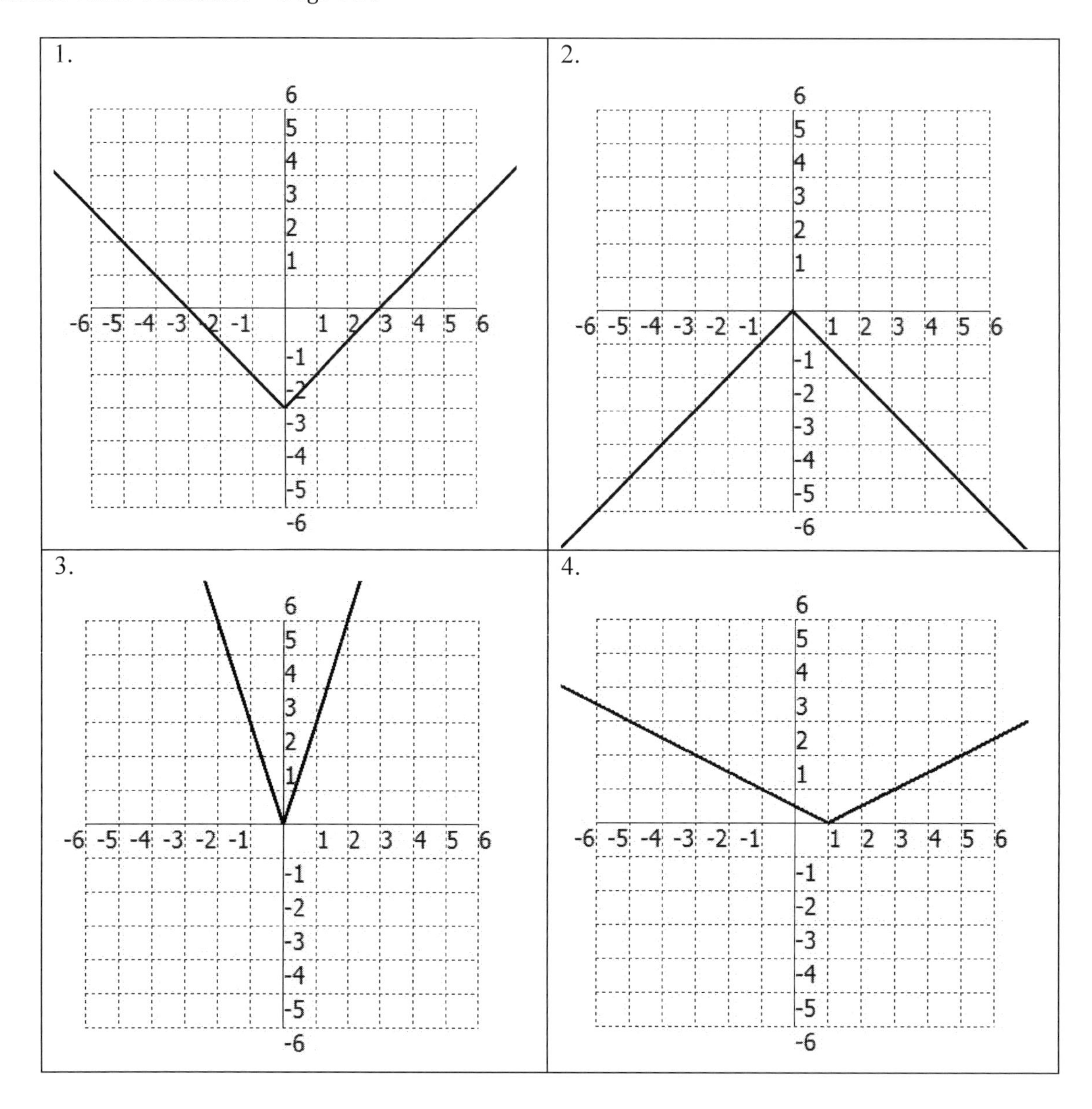

Identifying Families of Functions ~ Page 388

1. absolute value	2. exponential
3. quadratic	4. exponential
5. quadratic and linear	5. exponential and linear

Cubic, Square Root and Cube Root Functions ~ Page 393

No Practice Problems

Transformations of Functions ~ Page 396

1. (d)	2. (b)
3. (c)	4. (a)
5. $y=\frac{1}{2}x^2$; wider	6. $y=\lvert x\rvert-1$
7. $y=\lvert x+4\rvert$	8. $y=-(x-1)^2$

Piecewise-Defined Functions ~ Page 405

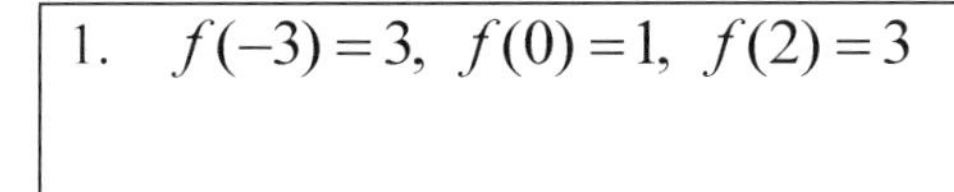 1. $f(-3)=3,\ f(0)=1,\ f(2)=3$	2. 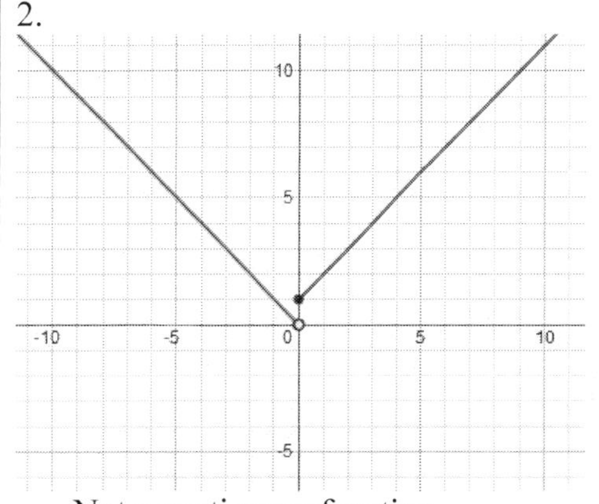 Not a continuous function.
 3. 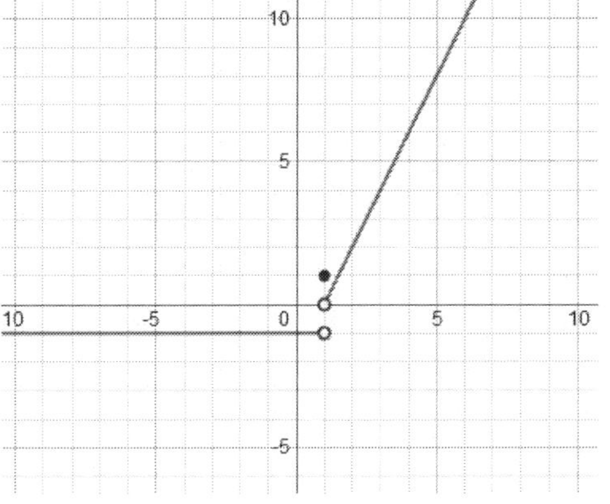	4. 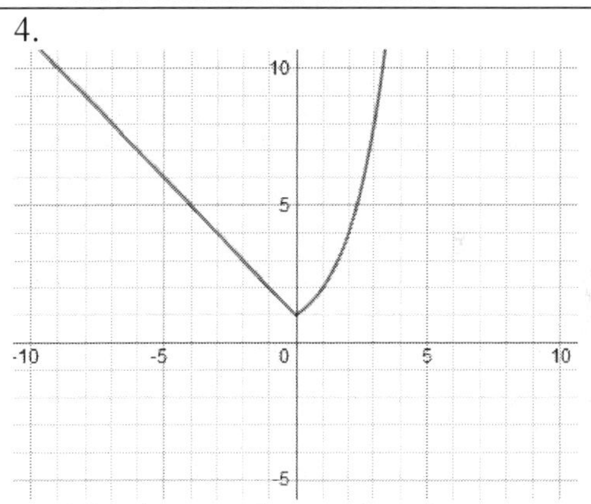 This is a continuous function.
5. 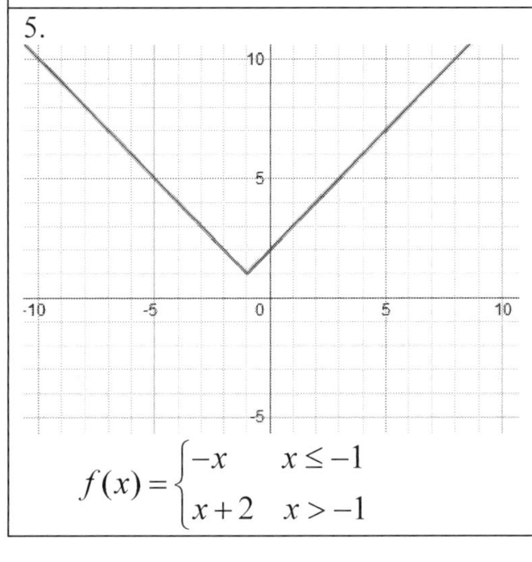 $f(x)=\begin{cases}-x & x\le -1\\ x+2 & x>-1\end{cases}$	6. $c(t)=\begin{cases}4t & 0<t\le 2\\ 2(t-2)+8 & 2<t\le 6\\ 16 & 6<t\le 8\end{cases}$

Step Functions ~ Page 408

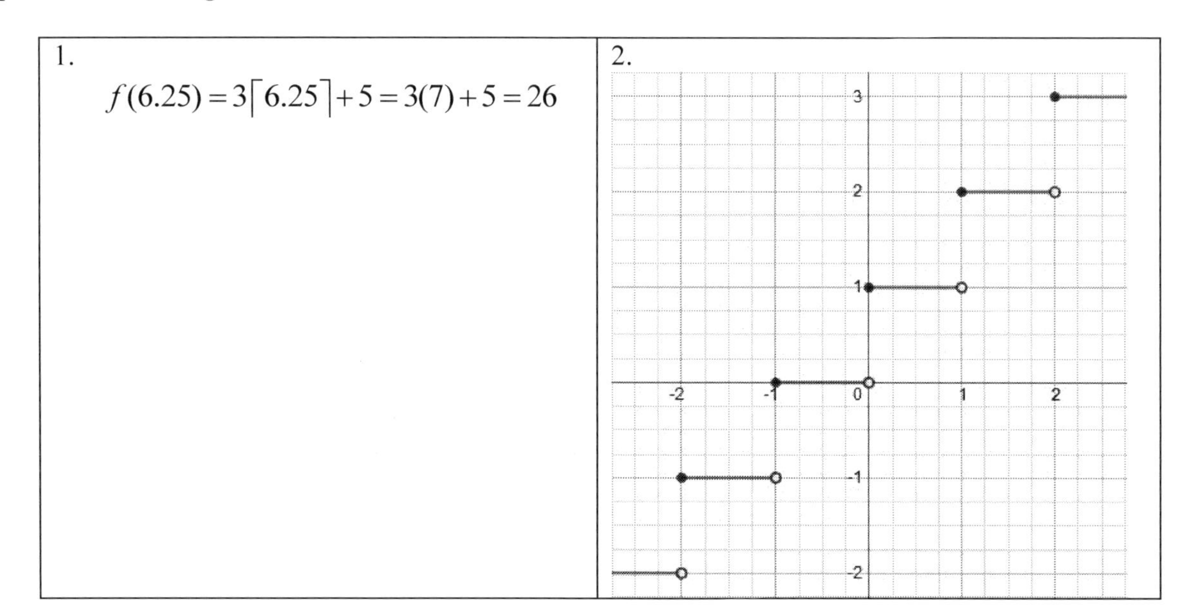

Regents Questions

NOTATION

A code next to each question number states from which exam the question came; for example, **IA AUG '09 [37]**, means the question appeared on the August 2009 Integrated Algebra Regents as question 37. Regents exam formats are abbreviated as follows:

CC = Algebra I Common Core, **IA** = Integrated Algebra, **A2** = Algebra 2 / Trigonometry,
MA = Math A, **MB** = Math B, **S1** = Sequential 1, **S3** = Sequential 3
9Y = Ninth Year Math, **EY** = Eleventh Year Math, **AL** = Algebra

Page 35 ~ Properties

1) 9Y JAN '72 [27] Ans: 4
2) 9Y JUN '75 [27] Ans: 1
3) 9Y JUN '78 [30] Ans: 2
4) IA FALL '07 [5] Ans: 3
5) IA AUG '08 [2] Ans: 2
6) IA JUN '09 [26] Ans: 3
7) IA JAN '11 [14] Ans: 4
8) IA JAN '12 [24] Ans: 3
9) IA AUG '12 [9] Ans: 1
10) IA AUG '13 [19] Ans: 1
11) IA JAN '14 [28] Ans: 3
12) CC JUN '14 [1] Ans: 1
13) 9Y JAN '71 [8] Ans: $\frac{1}{3x}$
14) IA JUN '10 [32] Ans: $-6a+42$, distributive

Page 40 ~ Solving Linear Equations in One Variable

1) IA AUG '08 [1] Ans: 4
2) IA JAN '10 [9] Ans: 2
3) IA JAN '11 [6] Ans: 1
4) IA AUG '12 [3] Ans: 3
5) IA JAN '13 [5] Ans: 4
6) IA JUN '14 [5] Ans: 1
7) IA JUN '14 [25] Ans: 2
8) 9Y JUN '75 [2] Ans: 3
9) IA FALL '07 [32] Ans: 4
10) IA JUN '11 [32] Ans: distributive, commutative
11) IA JUN '12 [38] Ans: 4

Page 46 ~ Solving Linear Inequalities in One Variable

1) 9Y JUN '69 [21] Ans: 2
2) MA JUN '00 [1] Ans: 2
3) MA AUG '04 [11] Ans: 4
4) IA FALL '07 [24] Ans: 1
5) IA AUG '08 [5] Ans: 4
6) IA JUN '09 [14] Ans: 1
7) IA AUG '09 [13] Ans: 4
8) IA AUG '10 [2] Ans: 1
9) IA AUG '11 [21] Ans: 4
10) IA AUG '13 [9] Ans: 1
11) IA AUG '13 [17] Ans: 3
12) IA JAN '14 [18] Ans: 1
13) IA JUN '14 [6] Ans: 1
14) IA JUN '10 [34] Ans: –12
15) IA JAN '12 [34] Ans: $x \geq 3$
16) IA JUN '13 [31] Ans: $x > 4$
17) CC JUN '14 [27] Ans: 2

Page 49 ~ Solving Proportions by Linear Equations

1) IA JAN '09 [6] Ans: 3
2) IA AUG '10 [12] Ans: 2
3) 9Y JUN '77 [8] Ans: 12
4) 9Y JUN '81 [32b] Ans: 5
5) 9Y JUN '82 [11] Ans: 18
6) 9Y AUG '83 [9] Ans: 8
7) 9Y AUG '85 [14] Ans: 12
8) 9Y JUN '87 [7] Ans: 7

Page 52 ~ Solving Equations with Fractions

1) 9Y APR '74 [30] Ans: 3
2) IA AUG '08 [20] Ans: 4
3) IA JUN '09 [7] Ans: 1
4) IA AUG '09 [9] Ans: 2
5) IA JUN '10 [19] Ans: 3
6) IA JAN '11 [12] Ans: 1
7) IA JAN '13 [28] Ans: 1
8) CC JUN '14 [5] Ans: 1
9) 9Y JAN '71 [3] Ans: 120
10) IA AUG '11 [39] Ans: 15

Page 56 ~ Literal Equations

1)	IA AUG '08 [8]	Ans: 3
2)	IA JAN '09 [11]	Ans: 2
3)	IA JUN '09 [13]	Ans: 3
4)	IA JAN '10 [16]	Ans: 4
5)	IA JUN '10 [23]	Ans: 2
6)	IA JAN '11 [25]	Ans: 4
7)	IA JAN '12 [28]	Ans: 1
8)	IA JUN '12 [15]	Ans: 1
9)	IA AUG '12 [30]	Ans: 3
10)	IA JUN '13 [16]	Ans: 1
11)	IA AUG '13 [30]	Ans: 2
12)	IA JAN '14 [25]	Ans: 1
13)	CC JUN '14 [23]	Ans: 1
14)	9Y JAN '71 [19]	Ans: $t=\frac{A-p}{pr}$
15)	MA JUN '99 [22]	Ans: $F=\frac{S+24}{3}$, 11.5
16)	IA AUG '11 [31]	Ans: $c=\frac{ab}{b+a}$

Page 64 ~ Translating Expressions

1)	9Y JUN '73 [28]	Ans: 4
2)	9Y JAN '71 [20]	Ans: 3
3)	9Y APR '74 [40]	Ans: 2
4)	IA FALL '07 [29]	Ans: 4
5)	IA JUN '08 [23]	Ans: 4
6)	IA JUN '10 [16]	Ans: 4
7)	IA AUG '10 [13]	Ans: 4
8)	IA JAN '11 [4]	Ans: 3
9)	IA JUN '11 [19]	Ans: 3
10)	IA AUG '11 [10]	Ans: 1
11)	IA JAN '12 [5]	Ans: 3
12)	IA JUN '12 [4]	Ans: 1
13)	IA AUG '12 [15]	Ans: 2
14)	IA JAN '13 [3]	Ans: 1
15)	IA JAN '13 [11]	Ans: 1
16)	IA JUN '13 [1]	Ans: 1
17)	IA JUN '13 [23]	Ans: 3
18)	IA JAN '14 [30]	Ans: 4
19)	9Y JAN '74 [16]	Ans: $25q+10d$
20)	MA SPR '98 [24]	Ans: $3d-1200$, \$1800
21)	MA JAN '02 [24]	Ans: $7x-2$

Page 68 ~ Translating "Each"

1)	IA JUN '09 [4]	Ans: 2
2)	CC JUN '14 [7]	Ans: 3
3)	CC JUN '14 [22]	Ans: 4

Page 70 ~ Translating Equations

1)	MA AUG '06 [27]	Ans: 4
2)	IA JAN '09 [15]	Ans: 2
3)	IA AUG '09 [1]	Ans: 2
4)	IA AUG '10 [11]	Ans: 4
5)	IA JUN '12 [25]	Ans: 3
6)	IA JAN '14 [13]	Ans: 3
7)	IA JUN '14 [18]	Ans: 1
8)	CC MAY '13 [4]	Ans: 4
9)	CC JUN '14 [16]	Ans: 2

Page 73 ~ Translating Inequalities

1)	IA FALL '07 [15]	Ans: 4
2)	IA JUN '08 [21]	Ans: 2
3)	IA AUG '08 [3]	Ans: 1
4)	IA JUN '09 [6]	Ans: 4
5)	IA JAN '10 [5]	Ans: 2
6)	IA AUG '11 [7]	Ans: 4
7)	IA AUG '12 [12]	Ans: 2
8)	IA JUN '13 [21]	Ans: 4
9)	IA JAN '14 [3]	Ans: 1

Page 77 ~ Word Problems – Linear Equations

1) MA JUN '04 [9] Ans: 2
2) MA JAN '07 [26] Ans: 4
3) MA JAN '08 [1] Ans: 4
4) IA JUN '08 [12] Ans: 3
5) IA AUG '08 [11] Ans: 2
6) IA JUN '09 [17] Ans: 1
7) IA AUG '09 [28] Ans: 4
8) IA JAN '11 [15] Ans: 2
9) IA JUN '11 [26] Ans: 1
10) IA AUG '11 [19] Ans: 1
11) IA AUG '12 [27] Ans: 2
12) 9Y JAN '75 [15] Ans: 10
13) 9Y JAN '87 [35] Ans: 20 2-cent and 30 22-cent stamps
14) MA JUN '99 [25] Ans: 38
15) MA AUG '00 [24] Ans: 21
16) MA JAN '07 [33] Ans: 6
17) IA JAN '12 [37] Ans: 7, 9, 11
18) IA AUG '13 [35] Ans: 36 and 64
19) CC MAY '13 [5] Ans: 2

Page 81 ~ Word Problems – Inequalities

1) IA JAN '09 [4] Ans: 1
2) IA AUG '09 [4] Ans: 1
3) IA JUN '12 [11] Ans: 3
4) IA AUG '12 [6] Ans: 4
5) IA FALL '07 [35] Ans: 7
6) IA JUN '08 [34] Ans: $10+2d \geq 75$, 33
7) IA JUN '11 [35] Ans: $0.65x+35 \leq 45$, 15

Page 84 ~ Determining Whether a Point is on a Line

1) 9Y JUN '75 [27] Ans: 4
2) MA AUG '06 [28] Ans: 1
3) IA JAN '10 [21] Ans: 1
4) IA JUN '10 [7] Ans: 3
5) IA AUG '10 [16] Ans: 4
6) IA JAN '12 [18] Ans: 4
7) IA AUG '12 [17] Ans: 4
8) IA JUN '13 [4] Ans: 3
9) 9Y JAN '78 [15] Ans: 2

Page 87 ~ Lines Parallel to Axes

1) IA AUG '08 [10] Ans: 2
2) IA AUG '09 [11] Ans: 1
3) IA AUG '10 [14] Ans: 2
4) IA JUN '11 [12] Ans: 4
5) IA JAN '13 [24] Ans: 3
6) IA JUN '13 [27] Ans: 2
7) IA JUN '14 [16] Ans: 1

Page 91 ~ Finding Slope Given Two Points

1) IA FALL '07 [16] Ans: 3
2) IA JUN '08 [20] Ans: 3
3) IA JAN '09 [13] Ans: 2
4) IA AUG '09 [15] Ans: 1
5) IA JAN '10 [7] Ans: 4
6) IA JUN '10 [4] Ans: 2
7) IA AUG '10 [5] Ans: 2
8) IA JUN '11 [10] Ans: 3
9) IA JAN '12 [15] Ans: 4
10) IA AUG '13 [10] Ans: 2
11) IA JUN '14 [10] Ans: 2

Page 95 ~ Finding Slope Given an Equation

1) IA JAN '11 [22] Ans: 2
2) IA JUN '12 [12] Ans: 4
3) IA JUN '13 [19] Ans: 4
4) IA JAN '14 [9] Ans: 2

Page 97 ~ Graphing a Linear Equation

1) IA JUN '12 [21] Ans: 4
2) CC JUN '14 [29] Ans: graph, no

Page 99 ~ Equations of Parallel Lines

1) 9Y APR '74 [35] Ans: 2
2) IA JUN '08 [14] Ans: 1
3) IA JAN '09 [26] Ans: 1
4) IA JAN '10 [26] Ans: 1
5) IA JUN '10 [15] Ans: 2
6) IA JAN '13 [22] Ans: 1
7) IA AUG '13 [7] Ans: 1

Page 101 ~ Writing a Linear Equation Given a Point and Slope

1) IA JUN '09 [22] Ans: 1
2) IA AUG '09 [27] Ans: 4
3) IA AUG '11 [8] Ans: 1
4) IA AUG '12 [19] Ans: 3
5) IA JAN '14 [6] Ans: 1
6) IA JUN '14 [19] Ans: 2
7) IA JAN '11 [34] Ans: $y = \frac{3}{4}x + 10$

Page 103 ~ Writing a Linear Equation Given Two Points

1) IA FALL '07 [13] Ans: 1
2) IA JAN '09 [10] Ans: 3
3) IA JAN '10 [13] Ans: 3
4) IA AUG '10 [29] Ans: 2
5) IA AUG '08 [36] Ans: $y = \frac{2}{5}x + 2$

Page 107 ~ Graphing Inequalities

1) IA FALL '07 [20] Ans: 2
2) IA JUN '09 [20] Ans: 1
3) IA JUN '10 [28] Ans: 4
4) IA JAN '12 [10] Ans: 1
5) IA JUN '13 [20] Ans: 4
6) IA AUG '13 [14] Ans: 2
7) IA JAN '14 [22] Ans: 3
8) IA JAN '10 [38] Ans: graph, yes, $4(1) - 3(-3) > 9$

Page 115 ~ Solving Systems of Equations Algebraically

1)	MA AUG ’00 [13]	Ans: 4
2)	IA JUN ’09 [25]	Ans: 2
3)	IA AUG ’09 [20]	Ans: 1
4)	IA JUN ’10 [12]	Ans: 3
5)	IA AUG ’10 [21]	Ans: 2
6)	IA AUG ’11 [9]	Ans: 3
7)	IA AUG ’13 [15]	Ans: 1
8)	IA JAN '14 [10]	Ans: 2
9)	IA JUN '14 [9]	Ans: 2
10)	CC JUN '14 [14]	Ans: 2
11)	9Y JAN ’70 [13]	Ans: 3
12)	9Y JUN ’76 [34]	Ans: (4,5)
13)	S3 AUG ’97 [17]	Ans: 2
14)	S3 AUG ’98 [37]	Ans: (3,–4)
15)	S3 JAN ’99 [41]	Ans: (2,7)
16)	S3 AUG ’99 [41]	Ans: (5,–3)
17)	IA JAN ’09 [37]	Ans: (–2,5)
18)	IA JUN ’12 [31]	Ans: 2

Page 120 ~ Solving Systems of Equations Graphically

1)	9Y JAN ’73 [21]	Ans: 1
2)	IA AUG ’12 [1]	Ans: 3
3)	IA JAN ’13 [4]	Ans: 3
4)	IA AUG ’09 [38]	Ans: (1,–3) w/ graph
5)	IA JAN ’12 [35]	Ans: (1,3) w/ graph

Page 124 ~ Solving Systems of Inequalities Graphically

1)	CC JUN '14 [4]	Ans: 2
2)	9Y JUN ’77 [31b]	Ans: graph, with A label in leftmost section
3)	IA JAN ’09 [38]	Ans: graph, infinitely many solutions such as (3,–3)
4)	IA AUG ’10 [37]	Ans: graph, infinitely many solutions such as (6, 6)
5)	IA JAN ’11 [39]	Ans: graph, with S label in rightmost section
6)	IA JUN ’11 [39]	Ans: graph, infinitely many solutions such as (–1,–1)
7)	IA AUG ’12 [39]	Ans: graph, infinitely many solutions such as (2,4)
8)	IA JUN '14 [38]	Ans: graph, with label *S* in lower left section

Page 131 ~ Solution Sets of Systems of Inequalities

1) IA AUG '08 [25] Ans: 4
2) IA JAN '10 [23] Ans: 2
3) IA JUN '10 [10] Ans: 1
4) IA AUG '11 [27] Ans: 2
5) IA JUN '12 [22] Ans: 4
6) IA JAN '13 [23] Ans: 2

Page 135 ~ Word Problems – Systems of Linear Equations

1) IA FALL '07 [8] Ans: 3
2) IA JUN '08 [6] Ans: 2
3) IA JUN '09 [12] Ans: 2
4) IA JAN '10 [3] Ans: 1
5) 9Y JAN '71 [33] Ans: 15
6) 9Y AUG '71 [36] Ans: 2
7) EY JAN '90 [9] Ans: 15
8) IA AUG '08 [37] Ans: $0.50 per marker, $0.15 per pencil
9) IA JAN '13 [35] Ans: $2.50 per notebook, $0.25 per pencil
10) IA JUN '14 [37] Ans: 64 apples and 44 oranges
11) CC JUN '14 [36] Ans: $2.35c+5.50d=89.50$, no, 10

Page 138 ~ Word Problems – Systems of Inequalities

1) CC MAY '13 [6] Ans: $x+y\leq 800$, $6x+9y\geq 5000$; yes w/ justification
2) MA JAN '02 [34] Ans: $x\leq 10$, $y\leq 12$ and $x+y\leq 16$ graphed with solution set shaded

Page 141 ~ Adding and Subtracting Polynomials

1) MA JAN '07 [7] Ans: 1
2) MA JUN '08 [5] Ans: 4
3) MA AUG '08 [7] Ans: 2
4) IA AUG '08 [19] Ans: 3
5) IA JUN '09 [23] Ans: 2
6) IA JUN '10 [3] Ans: 3
7) IA JAN '11 [26] Ans: 1
8) IA JUN '11 [30] Ans: 4
9) IA JAN '12 [13] Ans: 1
10) IA JUN '12 [26] Ans: 4
11) IA AUG '12 [5] Ans: 2
12) A2 JAN '13 [14] Ans: 1
13) IA JUN '13 [22] Ans: 1
14) IA AUG '13 [2] Ans: 1
15) CC JUN '14 [3] Ans: 2
16) IA JAN '14 [29] Ans: 4
17) IA JUN '14 [14] Ans: 2
18) 9Y JAN '70 [17] Ans: $-3x^2 - 11x$
19) 9Y JAN '73 [17] Ans: $11x - 4$
20) MA AUG '01 [23] Ans: $4x^2 + 10x + 2$
21) MA JAN '09 [34] Ans: $4x^2 + 8x - 10$

Page 147 ~ Multiplying Polynomials

1) 9Y JAN '71 [28] Ans: 4
2) 9Y JAN '76 [26] Ans: 3
3) S3 AUG '89 [17] Ans: 1
4) IA JUN '08 [7] Ans: 1
5) IA JAN '09 [24] Ans: 4
6) IA AUG '11 [14] Ans: 4
7) IA JUN '12 [10] Ans: 3
8) 9Y JUN '69 [18] Ans: $1 - x^4$
9) 9Y JAN '70 [15] Ans: $x^2 - 2x - 3$

Page 149 ~ Dividing a Polynomial by a Monomial

1) IA JAN '10 [11] Ans: 3
2) IA JUN '12 [3] Ans: 4
3) IA AUG '12 [22] Ans: 4
4) IA JAN '13 [16] Ans: 2
5) IA JAN '14 [12] Ans: 4
6) IA AUG '10 [31] Ans: $3a^2b^2 - 6a$

Page 153 ~ Irrational Numbers

1) CC JUN '14 [13] Ans: 3

Page 156 ~ Simplifying Radicals

1) IA JUN '08 [28] Ans: 1
2) IA JAN '09 [20] Ans: 3
3) IA JUN '09 [10] Ans: 2
4) IA AUG '09 [22] Ans: 2
5) IA JUN '11 [6] Ans: 3
6) IA JAN '12 [3] Ans: 3
7) IA FALL '07 [31] Ans: $30\sqrt{2}$
8) IA AUG '10 [33] Ans: $-12\sqrt{3}$
9) IA JAN '13 [31] Ans: $20\sqrt{3}$
10) IA AUG '13 [32] Ans: $12\sqrt{3}$

Page 160 ~ Operations with Radicals

1) 9Y JAN '73 [29] Ans: 4
2) 9Y JAN '74 [28] Ans: 2
3) IA JAN '10 [24] Ans: 4
4) IA JUN '10 [8] Ans: 3
5) IA JAN '11 [21] Ans: 3
6) A2 AUG '10 [1] Ans: 4
7) IA AUG '08 [34] Ans: $60-42\sqrt{5}$
8) IA AUG '11 [36] Ans: $-2\sqrt{3}$
9) IA JUN '12 [36] Ans: $6\sqrt{3}$
10) IA JUN '13 [36] Ans: $11+\sqrt{3}$
11) IA JAN '14 [31] Ans: $\sqrt{7}$
12) IA JUN '14 [36] Ans: $189\sqrt{2}$

Page 164 ~ Qualitative and Quantitative Data

1) IA JUN ’08 [19] Ans: 3
2) IA JUN ’09 [5] Ans: 3
3) IA JAN ’11 [16] Ans: 4
4) IA AUG ’11 [22] Ans: 4
5) IA JAN ’12 [11] Ans: 2
6) IA AUG ’12 [13] Ans: 3
7) IA JUN ’13 [8] Ans: 1
8) IA AUG ’13 [13] Ans: 3
9) IA JAN '14 [14] Ans: 3
10) IA JUN '14 [2] Ans: 3

Page 168 ~ Univariate and Bivariate Data

1) IA FALL ’07 [14] Ans: 2
2) IA JAN ’10 [14] Ans: 3
3) IA JUN ’10 [11] Ans: 3
4) IA JUN ’12 [6] Ans: 3

Page 177 ~ Frequency Tables and Histograms

1) IA JUN ’08 [22] Ans: 3
2) IA AUG ’08 [38] Ans: frequencies of 3, 7, 7, 3 w/ tallies, cumulative frequencies of 3, 10, 17, 20 w/ histogram
3) IA JUN ’09 [38] Ans: frequencies of 4, 5, 4, 8, 7, 2 w/ tallies and histogram
4) IA JUN ’10 [38] Ans: 30, 20, 71-80, 81-90 and 91-100
5) IA JAN ’11 [35] Ans: frequencies of 2, 2, 4, 6, 4 w/ tallies and histogram
6) IA AUG ’11 [32] Ans: histogram
7) IA AUG ’12 [34] Ans: 3, 0, 20
8) IA JAN '14 [32] Ans: 71-80

Page 186 ~ Central Tendency

1) S3 JUN '84 [32] Ans: 2
2) MA JAN '06 [18] Ans: 3
3) IA AUG '08 [4] Ans: 3
4) IA JAN '09 [7] Ans: 4
5) IA AUG '10 [20] Ans: 4
6) IA JAN '11 [18] Ans: 1
7) IA JAN '13 [29] Ans: 2
8) IA AUG '13 [27] Ans: 2
9) S3 AUG '90 [5] Ans: 6
10) MB JAN '03 [21] Ans: an outlier such as a very low score could greatly affect the range without affecting the median
11) IA FALL '07 [37] Ans: 225000, 175000, the median is closer to more values
12) IA JUN '08 [39] Ans: 315000, 180000, the median is closer to more values
13) IA JAN '10 [35] Ans: 81.3, 80, both increase by 5
14) IA JUN '11 [34] Ans: 12, 7, both increase by 3

Page 193 ~ Standard Deviation

1) MB AUG '08 [2] Ans: 2
2) MB JAN '04 [6] Ans: 1
3) MB JUN '02 [21] Ans: Adams, smaller SD means more consistent
4) S3 JUN '97 [39] Ans: 3.06
5) S3 JUN '00 [38] Ans: 16.2
6) S3 AUG '97 [42] Ans: 11.198
7) S3 JUN '98 [39] Ans: 5.48

Page 196 ~ Percentiles

1) IA JUN '12 [7] Ans: 4

Page 199 ~ Quartiles

1) IA JUN '10 [17] Ans: 3
2) IA JUN '12 [30] Ans: 3
3) CC JUN '14 [19] Ans: 3
4) A2 JAN '13 [31] Ans: 7

Page 205 ~ Box Plots

1) IA FALL '07 [9] Ans: 2
2) IA AUG '08 [18] Ans: 3
3) IA JAN '09 [29] Ans: 4
4) IA JUN '09 [15] Ans: 3
5) IA JAN '10 [1] Ans: 1
6) IA JAN '11 [13] Ans: 3
7) IA AUG '11 [6] Ans: 2
8) IA JAN '12 [20] Ans: 3
9) IA JUN '13 [14] Ans: 2
10) IA AUG '13 [12] Ans: 4
11) IA JAN '14 [8] Ans: 3
12) IA AUG '09 [39] Ans: plot ranging from 66 to 95 with quartiles at 72.5, 85, 90
13) IA AUG '10 [34] Ans: 120, 145, 292, 407, 452, plot
14) IA JAN '13 [37] Ans: plot, 3
15) IA JUN '14 [39] Ans: 8, 20, 32, 36, and 40; plot
16) CC JUN '14 [32] Ans: plot

Page 217 ~ Scatter Plots

1) IA FALL '07 [1] Ans: 2
2) IA AUG '10 [1] Ans: 3
3) IA JUN '11 [15] Ans: 2

Page 222 ~ Correlation and Causality

1) IA FALL '07 [7] Ans: 1
2) IA AUG '09 [8] Ans: 3
3) IA JAN '10 [30] Ans: 3
4) IA AUG '10 [17] Ans: 3
5) IA JUN '11 [22] Ans: 2
6) IA AUG '11 [4] Ans: 2
7) IA AUG '13 [1] Ans: 1
8) IA JUN '14 [27] Ans: 2

Page 226 ~ Identifying Correlation in Scatter Plots

1) IA JUN '08 [5] Ans: 4
2) IA JAN '10 [19] Ans: 2
3) IA JAN '11 [3] Ans: 3
4) IA AUG '11 [2] Ans: 1
5) IA JUN '12 [5] Ans: 2
6) IA AUG '12 [4] Ans: 1
7) IA JAN '13 [1] Ans: 1

Page 236 ~ Lines of Fit

1) IA AUG '08 [22] Ans: 4
2) IA AUG '09 [30] Ans: 2
3) IA JAN '12 [29] Ans: 4
4) IA AUG '12 [8] Ans: 3
5) IA JUN '13 [3] Ans: 3
6) IA JAN '14 [11] Ans: 2
7) MB JUN '07 [22] Ans: $y = 1.08x - 2125$
8) MB JUN '10 [27] Ans: $y = -0.112x + 23.448$, –5
9) MB JAN '03 [28] Ans: $y = 0.8345x + 14.6496$, 80
10) IA JUN '09 [36] Ans: graph, including line such as $y = 5x + 25$
11) IA JUN '10 [36] Ans: graph of line, no, line should cross near (18,13)

Page 245 ~ Residuals and Correlation Coefficients

1) A2 JUN '13 [16] Ans: 1
2) A2 JUN '10 [21] Ans: 2
3) MB JUN '07 [5] Ans: 4
4) MB AUG '03 [6] Ans: 4
5) A2 JAN '13 [3] Ans: 2
6) MB JUN '01 [9] Ans: 1
7) A2 JUN '12 [25] Ans: 1
8) CC JUN '14 [11] Ans: 3
9) MB JAN '02 [34] Ans: $y = 0.62x + 29.18$, 0.92, 83, w/plot

Page 253 ~ Determining if Relations are Functions

1) MB AUG '04 [3] Ans: 1
2) MB JUN '07 [15] Ans: 2
3) IA AUG '09 [19] Ans: 3
4) IA JAN '10 [18] Ans: 4
5) IA JAN '11 [5] Ans: 4
6) IA JUN '11 [16] Ans: 2
7) IA JAN '14 [5] Ans: 4
8) IA JUN '14 [13] Ans: 1

Page 256 ~ Determining if Graphs Represent Functions

1) IA FALL '07 [30] Ans: 4
2) IA JAN '09 [30] Ans: 4
3) IA JUN '09 [19] Ans: 3
4) IA JUN '10 [13] Ans: 4
5) IA JAN '12 [4] Ans: 3
6) IA JUN '12 [9] Ans: 1
7) IA JAN '13 [9] Ans: 3
8) IA AUG '13 [8] Ans: 3

Page 262 ~ Function Notation, Domain and Range

1) CC JUN '14 [2] Ans: 4
2) CC JUN '14 [17] Ans: 4
3) CC JUN '14 [30] Ans: yes, w/ justification

Page 265 ~ Function Graphs

1) CC JUN '14 [9] Ans: 3
2) CC JUN '14 [20] Ans: 1

Page 268 ~ Evaluating Functions

1) S3 JAN '89 [15] Ans: 3
2) S3 JAN '90 [20] Ans: 4
3) S3 AUG '94 [22] Ans: 2
4) IA JUN '14 [22] Ans: 4
5) EY JUN '85 [1] Ans: 16
6) S3 AUG '85 [1] Ans: 11
7) EY AUG '86 [6] Ans: 9
8) S3 AUG '00 [1] Ans: 1
9) CC MAY '13 [8] Ans: $A(x)=1.50x+6$ and $B(x)=2x+2.50$, 7, B

Page 272 ~ Rate of Change for Linear Functions

1) EY JUN '80 [17] Ans: 1
2) IA AUG '08 [7] Ans: 2
3) IA AUG '08 [23] Ans: 2
4) IA AUG '11 [15] Ans: 1
5) IA AUG '12 [23] Ans: 2

Page 276 ~ Average Rate of Change

1) IA JAN '09 [5] Ans: 1
2) CC MAY '13 [1] Ans: 4
3) CC JUN '14 [18] Ans: 1

Page 283 ~ Word Problems – Function Graphs

1) MA JAN '04 [12] Ans: 2
2) MA AUG '04 [10] Ans: 2
3) MA JAN '01 [21] Ans: B, 5 mins
4) CC MAY '13 [7] Ans: graph, 3.5

Page 287 ~ Percent Change

1) 9Y APR '74 [23] Ans: 2
2) MA JAN '03 [22] Ans: 43%
3) IA JUN '08 [35] Ans: $\frac{1}{6}$, 16.67%, $13.50
4) IA AUG '09 [35] Ans: 30.4%, no, 23.3%

Page 290 ~ Exponential Growth and Decay

1) IA FALL '07 [19] Ans: 3
2) IA JUN '08 [30] Ans: 2
3) IA JAN '09 [8] Ans: 4
4) IA AUG '09 [29] Ans: 3
5) IA JAN '10 [6] Ans: 2
6) IA JUN '10 [30] Ans: 1
7) IA JUN '11 [24] Ans: 2
8) IA AUG '11 [24] Ans: 2
9) IA JAN '12 [2] Ans: 1
10) IA JUN '12 [29] Ans: 2
11) IA AUG '12 [11] Ans: 3
12) IA JAN '13 [10] Ans: 3
13) IA JAN '14 [23] Ans: 4
14) IA JUN '09 [35] Ans: $5,583.86
15) IA JAN '11 [38] Ans: $24,435.19
16) IA AUG '13 [33] Ans: $2295
17) IA JAN '14 [33] Ans: $1159.27
18) CC JUN '14 [26] Ans: rate of decay, number of milligrams at the start

Page 297 ~ Exponential Functions

1) S3 JUN '84 [30] Ans: 2
2) S3 AUG '84 [34] Ans: 2
3) A2 JAN '11 [19] Ans: 3
4) IA AUG '08 [35] Ans: graph, no, $2^x > 0$ for all values of x
5) IA AUG '12 [33] Ans: graph

Page 302 ~ Sequences

1) MA AUG '07 [5] Ans: 2
2) A2 JUN '10 [1] Ans: 3
3) A2 AUG '10 [14] Ans: 1
4) A2 JAN '11 [5] Ans: 3
5) A2 JAN '12 [17] Ans: 4
6) A2 JAN '13 [4] Ans: 3
7) CC JUN '14 [21] Ans: 4
8) CC JUN '14 [24] Ans: 2
9) A2 FALL '09 [34] Ans: –3, –5, –8, –12
10) A2 JUN '12 [33] Ans: 9

Page 306 ~ Comparing Linear and Exponential Functions

1) CC JUN '14 [6] Ans: 4
2) CC JUN '14 [15] Ans: 3
3) CC JUN '14 [35] Ans: $A(n) = 175 - 2.75n$, 63

Page 309 ~ Factoring Out the Greatest Common Factor

1) MA JUN '04 [21] Ans: 3
2) IA JAN '14 [2] Ans: 3

Page 312 ~ Factoring a Trinomial

1) MA JAN '03 [18] Ans: 3
2) MA JAN '08 [14] Ans: 2
3) IA JUN '11 [5] Ans: 2
4) MA JAN '00 [4] Ans: 1
5) MA JUN '02 [6] Ans: 4
6) MA SPR '98 [6] Ans: 4
7) S3 JAN '00 [7] Ans: 7

Page 315 ~ Factoring the Difference of Two Perfect Squares

1) IA FALL '07 [6] Ans: 3
2) IA JUN '08 [4] Ans: 1
3) IA JAN '09 [9] Ans: 2
4) IA AUG '09 [2] Ans: 1
5) IA JAN '10 [22] Ans: 2
6) IA AUG '10 [8] Ans: 3
7) IA JUN '11 [1] Ans: 3
8) IA JAN '12 [1] Ans: 2
9) IA AUG '12 [7] Ans: 3
10) IA JAN '13 [6] Ans: 1

Page 318 ~ Factoring Completely

1) 9Y JAN '80 [29] Ans: 1
2) IA AUG '08 [6] Ans: 2
3) A2 FALL '09 [17] Ans: 4
4) IA JUN '10 [27] Ans: 2
5) IA JAN '11 [8] Ans: 2
6) IA JUN '11 [5] Ans: 2
7) IA AUG '11 [29] Ans: 2
8) IA JUN '12 [27] Ans: 4
9) IA JUN '14 [21] Ans: 4
10) S3 JAN '89 [12] Ans: $x(x-3)(x+2)$
11) MA JUN '05 [35] Ans: $3(x+7)(x-2)$
12) IA JUN '09 [32] Ans: $4x(x+3)(x-3)$
13) IA JAN '13 [32] Ans: $5x(x+2)(x-6)$
14) CC JUN '14 [31] Ans: $(x^2+7)(x+1)(x-1)$

Page 321 ~ Factoring by Grouping

1) A2 JUN '12 [14] Ans: 2
2) A2 JAN '13 [17] Ans: 3

Page 324 ~ Factoring Trinomials with Leads Other than 1

1) A2 JUN '10 [8] Ans: 4
2) 9Y JAN '72 [7] Ans: $(2x-3)$
3) EY JUN '86 [8] Ans: $x(2x-1)(x-5)$
4) S3 JAN '01 [11] Ans: $t(3t-4)(t+3)$

Page 328 ~ Solving Quadratic Equations

1) IA JAN '09 [14] Ans: 3
2) IA JUN '09 [2] Ans: 4
3) IA AUG '09 [21] Ans: 3
4) IA AUG '11 [20] Ans: 2
5) IA JAN '12 [23] Ans: 1
6) IA JAN '13 [20] Ans: 4
7) IA JUN '13 [11] Ans: 2
8) IA JAN '14 [27] Ans: 4
9) 9Y JAN '70 [4] Ans: 9
10) 9Y JAN '74 [32b] Ans: 3, –1
11) IA JAN '10 [34] Ans: {–2,3}
12) IA AUG '10 [36] Ans: {–15,2}
13) CC JUN '14 [33] Ans: $m(x)=x^2+10x+16$, $\{-8,-2\}$

Page 331 ~ Finding Quadratic Equations from Given Roots

1) 9Y APR '74 [30] Ans: 3
2) MA AUG '08 [25] Ans: 1
3) MA JAN '09 [13] Ans: 1
4) IA JAN '11 [28] Ans: 2
5) IA JUN '13 [26] Ans: 2
6) CC MAY '13 [3] Ans: 2
7) CC JUN '14 [12] Ans: 3
8) AL JAN '90 [12] Ans: $x^2-2x-35=0$
9) AL NOV '92 [7] Ans: $x^2+x-12=0$

Page 333 ~ Undefined Expressions

1) IA FALL '07 [28] Ans: 1
2) IA JUN '08 [17] Ans: 3
3) IA JAN '09 [25] Ans: 2
4) IA JUN '09 [16] Ans: 4
5) IA AUG '09 [18] Ans: 1
6) IA JUN '10 [14] Ans: 3
7) IA JUN '11 [25] Ans: 4
8) IA JAN '12 [14] Ans: 1
9) IA AUG '12 [25] Ans: 3
10) IA JUN '13 [15] Ans: 1
11) IA AUG '13 [16] Ans: 4
12) IA JUN '14 [29] Ans: 2

Page 336 ~ Solving Proportions by Quadratic Equations

1) IA JUN '08 [26] Ans: 4
2) IA JAN '10 [28] Ans: 4
3) IA AUG '12 [26] Ans: 3
4) IA JUN '13 [17] Ans: 2
5) IA FALL '07 [39] Ans: {6,–2}
6) IA JAN '11 [36] Ans: {4,–5}
7) IA JAN '14 [38] Ans: {12,–2}

Page 341 ~ Completing the Square

1) A2 JAN '11 [16] Ans: 2
2) A2 JUN '11 [22] Ans: 2
3) CC JUN '14 [8] Ans: 2
4) A2 FALL '93 [36] Ans: $\{3+\sqrt{7}, 3-\sqrt{7}\}$

Page 345 ~ Quadratic Formula and the Discriminant

1) MB JUN '10 [10] Ans: 3
2) A2 AUG '10 [9] Ans: 3
3) A2 AUG '10 [16] Ans: 4
4) A2 JAN '11 [2] Ans: 3
5) A2 JAN '13 [23] Ans: 4
6) A2 JAN '14 [11] Ans: 2
7) A2 JUN '14 [23] Ans: 3
8) CC JUN '14 [10] Ans: 2
9) A2 JAN '13 [32] Ans: $\frac{1\pm\sqrt{19}}{6}$

Page 348 ~ Word Problems – Quadratic Equations

1) IA FALL '07 [26] Ans: 4
2) IA AUG '08 [17] Ans: 2
3) IA JUN '10 [20] Ans: 1
4) IA AUG '11 [16] Ans: 2
5) IA JAN '12 [8] Ans: 3
6) IA AUG '13 [4] Ans: 3
7) 9Y JUN '66 [34] Ans: 5 and 15
8) 9Y JAN '72 [33] Ans: $(s+3)(s-2)=s^2$, $s=6$
9) 9Y JUN '65 [37] Ans: (a) $(x+40)(x+60)$; (b) 2400, 60x, x2, 40x; (c) Both equal $x^2+100x+2400$
10) 9Y JAN '73 [32] Ans: 6 and 7
11) 9Y JAN '80 [35] Ans: 5 and 11
12) 9Y JAN '81 [35] Ans: 2, 3, and 4
13) 9Y JUN '84 [34] Ans: 7 and 9
14) MB AUG '06 [34] Ans: 120 because $L=120$ when $t=0$, 4.2 mins.
15) IA JUN '08 [37] Ans: $w(w+15)=54$, 3, 18
16) IA JAN '10 [39] Ans: 6, 8, 10
17) MB JUN '10 [30] Ans: 3.71 and 4.71
18) CC JUN '14 [34] Ans: $(2x+16)(2x+12)=396$, justification, 3 m

Page 353 ~ Finding Roots Given a Parabolic Graph

1) IA JUN '09 [24] Ans: 3
2) IA AUG '09 [16] Ans: 2
3) IA JAN '11 [11] Ans: 4
4) IA JUN '13 [6] Ans: 3
5) IA JUN '14 [4] Ans: 3
6) CC MAY '13 [2] Ans: 3

Page 358 ~ Finding Vertex and Axis of Symmetry Graphically

1) IA JUN '08 [11] Ans: 1
2) IA AUG '08 [13] Ans: 1
3) MB AUG '09 [12] Ans: 3
4) IA JAN '09 [16] Ans: 2
5) IA JAN '10 [15] Ans: 2
6) IA JUN '10 [5] Ans: 1
7) IA AUG '11 [11] Ans: 2
8) IA AUG '12 [14] Ans: 4
9) IA JUN '14 [20] Ans: 1
10) IA JUN '11 [33] Ans: $x=1$, (1,–5)

Page 366 ~ Finding Vertex and Axis of Symmetry Algebraically

1) IA JUN '09 [18] Ans: 1
2) IA AUG '10 [18] Ans: 3
3) IA JAN '11 [27] Ans: 1
4) IA JAN '12 [19] Ans: 1
5) IA JUN '12 [14] Ans: 3
6) IA JAN '13 [14] Ans: 3
7) IA JAN '14 [16] Ans: 1
8) IA AUG '09 [34] Ans: $x = -2$, (–2,11)

Page 371 ~ Graphing Parabolas

1) IA JUN '08 [36] Ans: {–1,3} w/ graph
2) IA JUN '12 [34] Ans: {–4,2} w/ graph

Page 374 ~ Solving Quadratic-Linear Systems Algebraically

1) A2 AUG '10 [15] Ans: 2
2) IA JUN '08 [10] Ans: 4
3) IA AUG '08 [12] Ans: 2
4) IA JAN '09 [22] Ans: 2
5) IA JUN '11 [18] Ans: 2
6) IA JUN '12 [13] Ans: 2
7) IA JUN '13 [30] Ans: 1
8) IA JUN '14 [34] Ans: (3,7) and (–3,–5)

Page 377 ~ Solving Quadratic-Linear Systems Graphically

1) IA JAN '10 [12] Ans: 2
2) IA AUG '10 [10] Ans: 1
3) IA JAN '11 [2] Ans: 4
4) IA JAN '12 [7] Ans: 1
5) IA FALL '07 [38] Ans: {(0,5),(4,–3)} w/ graph
6) IA AUG '08 [39] Ans: {(–4,–5),(1,0)} w/ graph
7) IA JUN '09 [39] Ans: {(–1,8),(5,–4)} w/ graph
8) IA JUN '10 [39] Ans: {(–4,12),(2,0)} w/ graph
9) IA AUG '11 [38] Ans: {(2,5),(5,2)} w/ graph
10) IA JAN '13 [39] Ans: {(0,4),(3,7)} w/ graph
11) IA AUG '13 [37] Ans: {(–2,3),(2,–5)} w/ graph
12) IA JAN '14 [37] Ans: graph, (–4,–5) and (2,7)
13) CC JUN '14 [37] Ans: graph, 3, site *A*

Page 386 ~ Absolute Value Functions

1)	IA AUG '09 [25]	Ans: 3
2)	IA JAN '11 [17]	Ans: 3
3)	IA JAN '13 [33]	Ans: graph

Page 390 ~ Identifying Families of Functions

1)	IA FALL '07 [17]	Ans: 4
2)	IA JUN '08 [1]	Ans: 1
3)	IA AUG '10 [25]	Ans: 4
4)	IA JUN '11 [11]	Ans: 4
5)	IA AUG '11 [18]	Ans: 3
6)	IA JUN '13 [18]	Ans: 3
7)	IA JUN '14 [23]	Ans: 2

Page 395 ~ Cubic, Square Root and Cube Root Functions

1)	MB JAN '05 [32]	Ans: 670, 12, w/graph
2)	CC JUN '14 [25]	Ans: graph

Page 399 ~ Transformations of Functions

1)	IA FALL '07 [22]	Ans: 4
2)	IA JUN '08 [29]	Ans: 4
3)	IA JAN '10 [17]	Ans: 3
4)	IA AUG '10 [15]	Ans: 1
5)	IA JUN '11 [13]	Ans: 2
6)	IA JAN '12 [6]	Ans: 4
7)	IA AUG '12 [18]	Ans: 2
8)	IA JAN '13 [30]	Ans: 2
9)	IA AUG '13 [22]	Ans: 4
10)	IA JUN '10 [35]	Ans: graph, becomes wider (horizontally stretches)
11)	IA AUG '11 [34]	Ans: graph, becomes narrower (vertically stretches)
12)	IA JAN '14 [34]	Ans: graph, becomes narrower (horizontally shrinks)
13)	CC JUN '14 [28]	Ans: (4,–1), function shifted 2 to the right

Sample Regents Exam

Part I

1. (4)
2. (4)
3. (4)
4. (3)
5. (2)
6. (3)
7. (1)
8. (4)
9. (1)
10. (4)
11. (1)
12. (2)
13. (2)
14. (2)
15. (2)
16. (1)
17. (3)
18. (1)
19. (2)
20. (4)
21. (2)
22. (3)
23. (3)
24. (3)

Part II

25.

$$bc + ac = ab$$
$$c(b + a) = ab$$
$$c = \frac{ab}{b + a}$$

26.

$$4x^2 + 12x + 9 =$$
$$4x^2 + 6x + 6x + 9 =$$
$$2x(2x + 3) + 3(2x + 3) =$$
$$(2x + 3)(2x + 3)$$

so, the square root is $2x + 3$

27.

b = number of bags of soil
p = number of plants
$p \geq 5$
$4b + 10p \leq 100$

28.

$y = 0.75(35) - 0.25 = 26$ points
$32 - 26 = 6$

29.

$$m = \frac{2000 - 0}{2 - 0} = 1000 \text{ cm/s}$$

30.

Line A. Most of the points are closer to Line A than to Line B.

31.

point B
5 minutes

32.

$Q_3 - Q_1 = 89 - 73 = 16$

Part III

33.

	Fiction	Nonfiction	Total
Hardcover	28	52	80
Paperback	84	36	120
Total	112	88	200

	Fiction	Nonfiction	Total
Hardcover	14%	26%	40%
Paperback	42%	18%	60%
Total	56%	44%	100%

34.

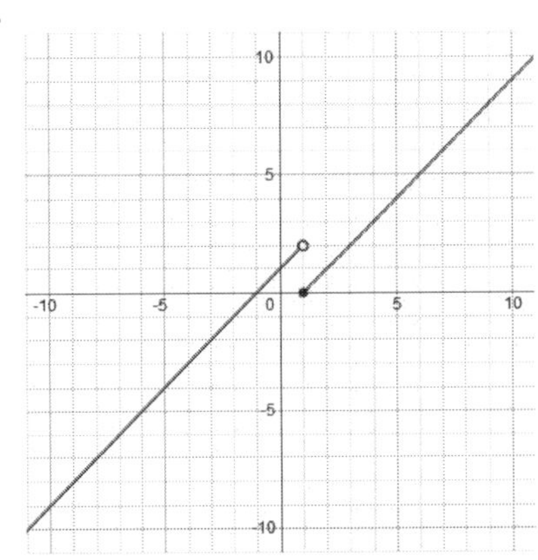

35.

120 liters; the y-intercept shows the number of liters at time 0

Method 1:

$$-5t^2-8t+120=0$$

$$t=\frac{-b\pm\sqrt{b^2-4ac}}{2a}=$$

$$\frac{8\pm\sqrt{(-8)^2-4(-5)(120)}}{2(-5)}=$$

$$\frac{8\pm\sqrt{2464}}{-10}\approx\frac{8\pm 49.64}{-10}$$

$$t=\frac{8-49.64}{-10}=4.164\approx 4.2$$

Method 2:

$$-5t^2-8t+120=0$$

$$25t^2+40t-600=0$$

$$25t^2+40t=600$$

$$25t^2+40t+16=600+16$$

$$\boxed{\frac{b^2}{4a}=\frac{1600}{100}=16}$$

$$(5t+4)^2=616$$

$$5t+4=\pm\sqrt{616}$$

$$t=\frac{-4\pm\sqrt{616}}{5}$$

$$t\approx 4.2$$

36.

Area of $ABCD=(x+40)(x+60)=x^2+100x+2400$

Sum of areas of I, II, III, and IV is $2400+60x+x^2+40x=x^2+100x+2400$

Part IV.

37.

a)

$$R(x)=500+20x$$

$$E(x)=6x$$

b)

$$P(x)=R(x)-E(x)=$$

$$500+20x-6x=500+14x$$

c)

$$500+14x\geq 2000$$

$$14x\geq 1500$$

$$x\geq 107.14...$$

The club needs to sell at least 108 calendars.

Made in the USA
Charleston, SC
10 February 2015